Energy Engineering - Applications and Batteries

Edited by Muhammad Sultan, Hong Zhao, Muhammad Wakil Shahzad, Zhaoli Zhang, Uzair Sajjad and Xiangkun Bo

Published in London, United Kingdom

Energy Engineering - Applications and Batteries
http://dx.doi.org/10.5772/intechopen.1008572
Edited by Muhammad Sultan, Hong Zhao, Muhammad Wakil Shahzad, Zhaoli Zhang, Uzair Sajjad and Xiangkun Bo

Contributors
Alexander Fandakov, Alireza Soleimani, Anna Pinnarelli, Brandon Kitch, Danko Brezak, Elijah Collins, Hafiz M.U. Raza, Hepsibah Priyadarshini Christopher, Johannes Werfel, Juraj Benić, Karsten Mueller, Michael Clauss, Muhammad Ahmad Jamil, Muhammad Ali Imran, Muhammad Asad Ali, Muhammad Hamid Mahmood, Muhammad Haseeb Siddiqui, Muhammad Sultan, Muhammad Wakil Shahzad, Mujtaba Sheikh, Nicholas Hoff, Oliver Nolte, Pasquale Vizza, Porter Little, Reece Durbin, Robert M. Stwalley III, Seshan Calapatti Suresh, Sudha Venkatachalam, Tovia Owens, Uzair Sajjad, Waheeduddin Hyder, Waqas Ahmed Siddique, Wenming Liao, Zhaoli Zhang

First published in London, United Kingdom, 2026 by IntechOpen
IntechOpen is the global imprint of INTECHOPEN LIMITED, registered in England and Wales, registration number: 11086078, 167-169 Great Portland Street, London, W1W 5PF, United Kingdom

For EU product safety concerns: IN TECH d.o.o., Prolaz Marije Krucifikse Kozulić 3, 51000 Rijeka, Croatia, info@intechopen.com or visit our website at intechopen.com.

British Library Cataloguing-in-Publication Data
A catalogue record for this book is available from the British Library

Energy Engineering - Applications and Batteries
Edited by Muhammad Sultan, Hong Zhao, Muhammad Wakil Shahzad, Zhaoli Zhang, Uzair Sajjad and Xiangkun Bo
p. cm.
Print ISBN 978-1-83634-653-1
Online ISBN 978-1-83634-652-4
eBook (PDF) ISBN 978-1-83634-654-8

If disposing of this product, please recycle the paper responsibly.

Meet the editors

Dr. Muhammad Sultan is an Associate Professor of Energy & Environmental Engineering at Bahauddin Zakariya University (Pakistan), and a globally recognized researcher. With a Ph.D. and postdoctoral research conducted in Japan and Canada, he has earned accolades as one of the World's Top 2% Scientists (Stanford University/Elsevier) and as the leading Pakistani scientist in Agricultural Engineering (AD Scientific Index). Dr. Sultan has authored over 210 journal articles, with a cumulative impact factor of over 1000. In addition, he published 20 books and 40 chapters, amassing over 9,000 citations, an h-index of 50, and an i10-index of 160. As an editor for 20 international journals, he conducts research on sustainable energy systems, advanced cooling technologies, and agricultural engineering innovations, contributing to both academic excellence and global sustainability efforts.

Dr. Hong Zhao is a Professor at Foshan University (Hydrogen Lab, School of Materials Science and Hydrogen Energy) in China, and a globally recognized researcher in advanced materials and energy storage. With postdoctoral training and research experience from the Hong Kong University of Science and Technology (HKUST), he has earned notable recognition, including the Special Awards at the 45th International Exhibition of Inventions in Geneva, Switzerland (2017) for leading the Flexible/Fast-charging Lithium-Ion Batteries Project. Dr. Zhao previously served as a Senior Engineer at the Nano and Advanced Materials R&D Institute Co., Ltd. (a subsidiary of HKUST) and as an Associate Researcher at HKUST. Currently, he holds multiple concurrent senior positions as Co-Founder of Ping Shan Technology and Technical and Product Consultant of Chung Pak Battery Works. His research spans hydrogen energy systems, advanced battery technologies, and materials science innovations, contributing to both academic excellence and global sustainable energy development.

Dr. Shahzad is working as a Professor in the Mechanical Engineering Department at Northumbria University (NU), Newcastle Upon Tyne, United Kingdom. He holds a Ph.D. in Mechanical Engineering from the National University of Singapore (2014) and a BSc from the University of Engineering and Technology Lahore, Pakistan (2007). He is a Chartered Engineer from the Engineering Council UK and a Fellow of AdvanceHE UK. Dr. Shahzad led the renewable energy and water generation research lab and education at Northumbria University. He conducts multidisciplinary research on the water-energy-environment nexus and trains the next generations of scientists capable of developing innovative solutions for sustainable water and cooling production, renewable energy harvesting, and smart grid integration and control strategy for process optimization.

Dr. Zhaoli Zhang is an associate professor at the School of Mechanical Engineering, Southwest Jiaotong University, China. He obtained his Bachelor's, Master's, and doctoral degrees in engineering from Southwest Jiaotong University. He once studied as a visiting scholar at Arizona State University in the USA. He also worked as a postdoctoral researcher at the Vrije Universiteit Brussel in Belgium. His research focuses on phase-change energy storage and energy-efficient utilization technologies, and he leads several projects funded by government and industry. To date, he has published over 20 SCI papers as first or corresponding author in international journals.

Dr. Uzair Sajjad is currently working as a postdoctoral fellow at the Interdisciplinary Research Center of Sustainable Energy Systems at King Fahd University of Petroleum and Minerals. Previously, he served as a Research Assistant Professor in the Department of Energy and Refrigerating Air-Conditioning Engineering at National Taipei University of Technology in Taiwan for 2. 5 years. He completed his Ph.D. at National Chiao Tung University in Taiwan and, for his research excellence, was awarded two science and technology awards by CTCI, Taiwan. He is conducting investigations on heat transfer, energy, applications of machine learning and artificial intelligence, surface engineering, heat pumps, vacuum freeze-drying, and additive manufacturing. He has won 3. 5 million NTD grants from the National Science and Technology Council (NSTC), Taiwan, China. The National Science and Technology Council of Taiwan has awarded him a project grant. In these areas, he has published more than 80 SCI articles and book chapters. In addition, he has served as a reviewer and guest editor for numerous well-reputed journals.

Dr. Bo Xiangkun is currently working as a Postdoctoral Research Fellow at the Institute of Textiles and Clothing, The Hong Kong Polytechnic University. He completed his Ph.D. in Mechanical Engineering from City University of Hong Kong in 2024. He previously obtained his Master's degree in Chemical Engineering from Nanjing University in 2017 and Bachelor's degree in Chemical Engineering and Technology from Qiqihar University in 2014. Prior to his doctoral studies, he worked as an engineer at Shenzhen Yanyitech New Material Co., Ltd. (2019-2020) and as an R&D engineer at Ningde Amperex Technology Limited (2017-2018). His research focuses on sustainable energy solutions, triboelectric nanogenerators, biomechanical energy harvesting, stretchable ionic conductors, and battery technologies. Dr. Bo has demonstrated exceptional research productivity, with more than 80 publications in high-impact journals, including Advanced Materials, Nano Energy, Chemical Engineering Journal, and Small, in which he frequently serves as the first author.

Contents

Preface

Energy engineering technologies and applications have become a cornerstone of modern technological development, underpinning efforts to improve efficiency, sustainability, and resilience across industrial, agricultural, and transportation systems. As global energy demand continues to rise amid concerns about climate change, resource depletion, and environmental impacts, innovative energy engineering solutions are essential to ensure a sustainable and secure energy future. This book focuses on energy engineering applications and advanced battery technologies, presenting contemporary research and practical insights that address the associated challenges through advanced engineering approaches and emerging technologies.

This book is organized into two sections, comprising eight chapters. The first section "Energy Engineering Applications" comprises Chapters 1 to 5, which focus on applied energy engineering solutions across thermal systems, hydraulic and mechanical systems, smart energy management, and sustainable propulsion technologies, covering topics such as hybrid indirect evaporative cooling for human, agricultural, and industrial applications, energy analysis of advanced hydraulic control methods, the integration of machine learning and IoT for energy efficiency and emission control, modern energy storage systems to address rising energy demands, and alternative propulsive and drivetrain methodologies for agricultural and off-road vehicles. The second section, "Energy Engineering for Batteries," comprises Chapters 6-8, which address advanced battery engineering topics, including computational modelling of electrode–electrolyte interfaces from lithium-ion to post–lithium-ion systems, emerging technologies for next-generation vehicle battery systems, and innovative template-based synthesis methods for bimetallic oxides used in lithium-ion batteries.

The opening chapter, "Evaluation of Hybrid Indirect Evaporative Cooling Systems for Human Thermal Comfort, Agricultural, and Industrial Applications", focuses on advanced cooling strategies to improve thermal comfort and reduce energy consumption. The chapter examines the performance of hybrid evaporative cooling systems and discusses their applicability across residential, agricultural, and industrial environments.

The second chapter, "Energy Analysis of Control Methods in a Dual-Pump Direct-Driven Hydraulic System", provides a detailed assessment of control strategies for hydraulic systems. It highlights how optimized control methods can significantly enhance energy efficiency, system reliability, and operational performance in industrial applications.

In the third chapter, "Green Energy 4.0: An Overview of the Synergy between Machine Learning and IoT for Energy Efficiency and Emission Control", the authors explore the role of digital technologies in modern energy systems. This chapter discusses how machine learning and Internet of Things-based solutions enable intelligent monitoring, predictive control, and emission reduction, thereby supporting the transition toward smarter, greener energy infrastructures.

The fourth chapter, "Energy Storage Systems: Bridging the Gap Between Supply and Growing Demand", addresses the growing need for efficient energy storage solutions to meet rising global energy demand. It discusses the existing and emerging storage technologies and their role in enhancing grid stability, energy flexibility, and sustainable power management.

The fifth chapter, "Alternative Propulsive and Drivetrain Methodologies for Energy Engineering in Agricultural and Off-Road Vehicles", examines innovative propulsion systems designed to improve energy efficiency and reduce environmental impact. The chapter highlights alternative drivetrain concepts, electrification strategies, and hybrid solutions tailored to agricultural and off-road applications.

The sixth chapter, "Computational Modelling of Electrode/Electrolyte Interfaces: Challenges and Perspectives from Li-Ion to Post Li-Ion Batteries", focuses on modelling approaches for understanding electrochemical interfaces, addressing current challenges and future directions for next-generation battery technologies.

In the seventh chapter, "Novel Technologies for the Next-Generation of Vehicle Battery Systems", readers are introduced to recent innovations aimed at improving battery performance, safety, and energy density for electric and hybrid vehicles. The chapter discusses technological advancements that support the growing adoption of electrified transportation systems.

The final chapter, "Template Methods for the Synthesis of Bimetallic Oxides Used in Lithium-Ion Batteries", provides insights into advanced material synthesis techniques. It emphasizes the importance of controlled material design for enhancing the electrochemical performance and durability of lithium-ion battery electrodes.

The book serves as a valuable reference for researchers, engineers, graduate students, and industry professionals engaged in energy engineering and/or the battery field. The volume contributes to ongoing efforts toward energy efficiency, decarbonization, and technological innovation.

As editors of this book, we express our sincere appreciation to all contributing authors for their high-quality research and dedication, to the reviewers for their constructive evaluations, and to the publishing team for their support in realizing this work. We hope this volume will inspire further research, collaboration, and innovation in energy engineering and advanced battery technologies.

Muhammad Sultan
Bahauddin Zakariya University,
Multan, Pakistan

Hong Zhao
Foshan University,
Foshan, China

Muhammad Wakil Shahzad
Northumbria University,
Newcastle upon Tyne, United Kingdom

Zhaoli Zhang
Southwest Jiaotong University,
Chengdu, China

Uzair Sajjad
King Fahd University of Petroleum and Minerals,
Saudi Arabia

Xiangkun Bo
The Hong Kong Polytechnic University,
Hong Kong, China

Section 1

Energy Engineering Applications

Chapter 1

Evaluation of Hybrid Indirect Evaporative Cooling Systems for Human Thermal Comfort, Agricultural, and Industrial Applications

Muhammad Ali Imran, Muhammad Sultan, Muhammad Wakil Shahzad, Muhammad Ahmad Jamil, Hafiz M.U. Raza, Muhammad Hamid Mahmood, Muhammad Asad Ali, Muhammad Haseeb Siddiqui, Uzair Sajjad and Zhaoli Zhang

Abstract

This study explores the applications of hybrid indirect evaporative cooling (IEC) systems as an energy-efficient and climate-resilient alternative approach to conventional systems and their viability under different climates, offering cooling solutions. This book chapter systematically reviews standalone and hybrid IEC systems that incorporate mechanical vapor compression (MVC), desiccant dehumidification, and other innovations to overcome the limitations of conventional EC systems, particularly in humid and hot varying conditions. In this regard, a wide spectrum of state-of-the-art applications is assessed, including human thermal comfort in various types of buildings, crops in greenhouses, comfort of livestock in shelter houses, fresh and dried food storage, and industrial processes, for example, artifact preservation, tobacco curing, rubber goods, and mobile systems. Explored case studies and their prospective performance display that hybrid IEC systems have great potential to enhance energy efficiency, performance, and environmental sustainability. This assessment offers critical insights for professionals and policymakers toward the expansion and implementation of hybrid IEC systems in numerous areas.

Keywords: temperature control, humidity control, evaporative cooling, heat pump systems, HVAC applications

1. Introduction

Global energy demand is drastically increasing due to both rising population growth and the frequency and intensity of heatwaves in response to climate change. It is estimated that global energy demand has increased by approximately 1.53 times compared to 1995. Consequently, the need for air conditioning has become inevitable in many regions [1, 2]. Currently, conventional mechanical vapor compression (MVC)-based systems dominate the market, but they are highly expensive and energy-intensive. Additionally, these systems utilize hazardous refrigerants that pose risks to environmental safety and contribute to carbon footprints [3]. It is of great importance to protect the environment and air quality by optimizing sustainable approaches [4]. Given these concerns, evaporative cooling (EC) systems have emerged as promising energy-efficient and environmentally friendly alternative solutions, including direct, indirect, and advanced indirect evaporative cooling (IEC) systems [5]. These energy-efficient systems reduce air temperature through the natural water evaporation process.

Despite the advantages, direct evaporative cooling (DEC) is ineffective for humid climates due to moisture addition. The adaptability of indirect IEC is advantageous over DEC; however, inconsistent cooling performance under different climatic conditions is the key concern [6]. These limitations are managed by hybrid systems, utilizing the strengths of EC systems in combination with MVC and other technologies [7]. These hybrid IEC systems provide high effectiveness and consistent performance, along with reduced energy consumption and environmental impact. These sustainable and efficient solutions have the potential to support the growing global cooling demand. The configuration and application of the hybrid systems are discussed in the following sections.

Temperature and humidity-controlled conditions are critically important for human thermal comfort, health, and performance. These systems are also engineered to maintain suitable climatic conditions tailored to the needs of various sectors [8, 9]. For instance, they ensure optimal environmental conditions in residential, commercial, and industrial buildings [10, 11]. Likewise, stable temperature and humidity-controlled environments are vital for patient recovery in hospitals [12, 13] and for preserving artifacts in museums [14]. In addition, such controlled temperature and humidity conditions are extended to different domains for both thermal comfort and technical requirements, such as automobiles [15], wet markets [16, 17], supermarkets [18], marine ships [19], industrial processes [20, 21], electronic cooling setups [22], and data centers [23]. The application of these controlled microclimatic conditions will be discussed in the following sections. Beyond these applications, the maintenance of microclimatic environments is integral across agricultural sectors. These energy-efficient systems are effective for improving agricultural product storage, such as perishable fruits and vegetables [24, 25], dried fruit storage [26], milk preservation [27], poultry and livestock [28], crop growth and yield in greenhouses [29], horticulture [30], and even in tractor cabins for thermal comfort [31, 32]. This book chapter discusses state-of-the-art applications of temperature and humidity-control systems, which are capable of reducing energy consumption and minimizing damage to the environment.

Given the rising global energy demands and the urgent need for climate-resilient cooling technologies, there is a critical requirement for systems that ensure optimal thermal conditions across diverse climatic zones while minimizing environmental impacts. Hybrid IEC systems have emerged as a promising solution, offering energy efficiency, environmental sustainability, and broad adaptability for various applications. This chapter provides a section-wise evaluation as follows:

- Section 2 introduces the fundamental principles of standalone and hybrid IEC systems, highlighting key mechanisms and configurations.
- Section 3.1 reviews human comfort applications in buildings, such as residences, hospitals, schools, and mosques, where the hybrid IEC systems have demonstrated enhanced indoor air quality and energy savings.
- Section 3.2 explores agricultural uses, including greenhouses, dairy farms, fruit and vegetable storage, and livestock environments, where controlled temperature and humidity are critical for productivity and product preservation.
- Section 3.3 examines industrial applications such as tobacco curing, matchstick manufacturing, rubber processing, mobile cold storage, and maritime cooling, showing the system's viability in thermally and humidity-sensitive industries.

Across these diverse use cases, the chapter presents a state-of-the-art assessment of the hybrid IEC systems, shedding light on system designs, performance improvements, energy-saving potential, and practical implementation. With growing global interest, these systems are poised to play a pivotal role in future sustainable cooling strategies.

2. Indirect evaporative cooling systems

Evaporative cooling represents the generation of a cooling effect due to water evaporation, as water molecules absorb heat from the surroundings and transform into a gaseous form. Generally, evaporative cooling systems can be classified into direct and indirect EC. In DEC systems, hot and dry ambient air directly interacts with the wet cooling pad, and water evaporates by absorbing heat from the air. Therefore, the air temperature decreases as water absorbs the sensible heat of the air, and the humidity of the air increases due to the conversion of water into vapor as latent heat. This cooled and humidified air is directly supplied to indoor space [6, 33]. It represents an isenthalpic process, wherein air enthalpy remains constant. The IEC system is a method that indirectly reduces the ambient air temperature without manipulating humidity. This system uses an indirect heat exchanger to transfer cooling produced via the evaporation process. The details of the IEC systems are discussed in the following section.

2.1 Standalone indirect evaporative cooling systems

IEC systems indirectly reduce the ambient air temperature without manipulating humidity. This system uses an indirect heat exchanger to transfer cooling produced via the evaporation process, avoiding excessive humidification of cooled air. The heat exchanger deals with two separate airstreams side by side, as shown in **Figures 1a** and **b**. In the IEC system, working air passes over a wet surface

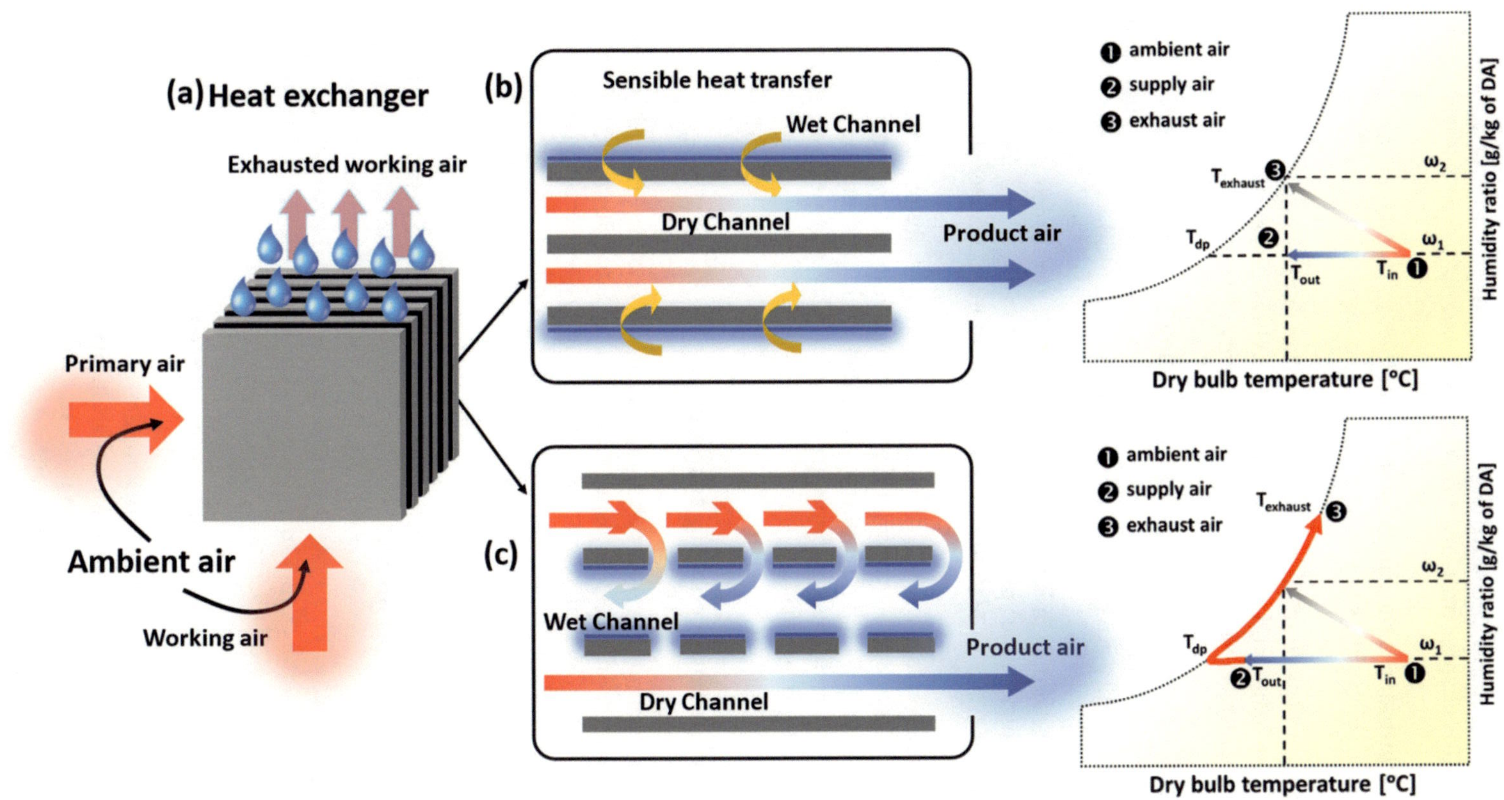

Figure 1.
Schematic representation of an indirect evaporative cooling system (IEC): (a) basic concept of the heat exchanger, and the working principle and thermodynamic illustration of (b) IEC and (c) advanced IEC systems.

side and cools down through the evaporation process, while sensibly absorbing heat from product air in the dry channel [34]. As per psychrometric representation, product air cools at a constant humidity, while working air humidity increases in terms of latent heat (**Figure 1b**). The IEC cooling cannot reduce below the wet bulb temperature, which varies from 40% to 80% under different ambient conditions [35].

Advanced IEC systems are a modified version of IEC that improve air cooling up to the dew point temperature [36]. In advanced IEC systems, a part of the air is diverted to the wet channels, where it absorbs moisture through evaporation and removes heat from the exchanger surfaces, as described in **Figure 1c**. This system maintains a large temperature difference between the plates, which cools the remaining air in the dry channels without increasing the humidity [10, 37, 38]. In psychrometric representations, the supply air flows toward the dew point, and the working air moves toward saturation. Studies have shown a wet-bulb effectiveness of more than 100% for advanced IEC systems [10, 39, 40]. Additionally, several studies have explored various standalone and integrated systems, such as IEC, IEC-DEC, and IEC-DEC [6, 41]. The case studies of these systems will be discussed in detail in the following sections.

2.2 Hybrid indirect evaporative cooling systems

Despite high energy efficiency, the IEC systems are restricted by inherent limitations such as unstable cooling efficiency under humid conditions and varying ambient conditions [7, 42]. To mitigate these concerns, a hybrid system couples the IEC system with mechanical vapor compression (MVC) systems. In hybrid IEC-MVC systems, energy-efficient IEC systems pre-cool the air, and MVC is used to further process the air to achieve stable performance, as shown in **Figure 2** [43]. Additionally, it recovers the cold energy of exhaust air by recirculating it to the wet channels. These hybrid systems not only enhance the overall system efficiency but also extend applicability across diverse climatic zones and have been explored in different configurations such as IEC-MVC [43], MVC-IEC [44], and IEC-DEC-MVC [45]. Therefore, hybrid IEC-MVC systems have received attention as an emerging solution toward a sustainable and climate-resilient cooling approach. The aim is to identify the most promising hybrid systems by evaluating their coefficient of performance (COP), energy-saving potential, practical applicability, and temperature reduction under dynamic climatic conditions. The performances of these systems will be discussed in the next section.

2.3 Comparisons of advantages and limitations

Table 1 provides an overview of the advantages and limitations associated with the different IEC-based systems. The standalone IEC and advanced systems are sustainable and require low-energy, while hybrid IEC-MVC systems allow more stable performance and support high cooling loads. The complexity, maintenance, and initial cost increase with the sophistication and applicability of the system. Meanwhile, the specific application requirements to control temperature and humidity, along with microclimatic conditions, require further consideration.

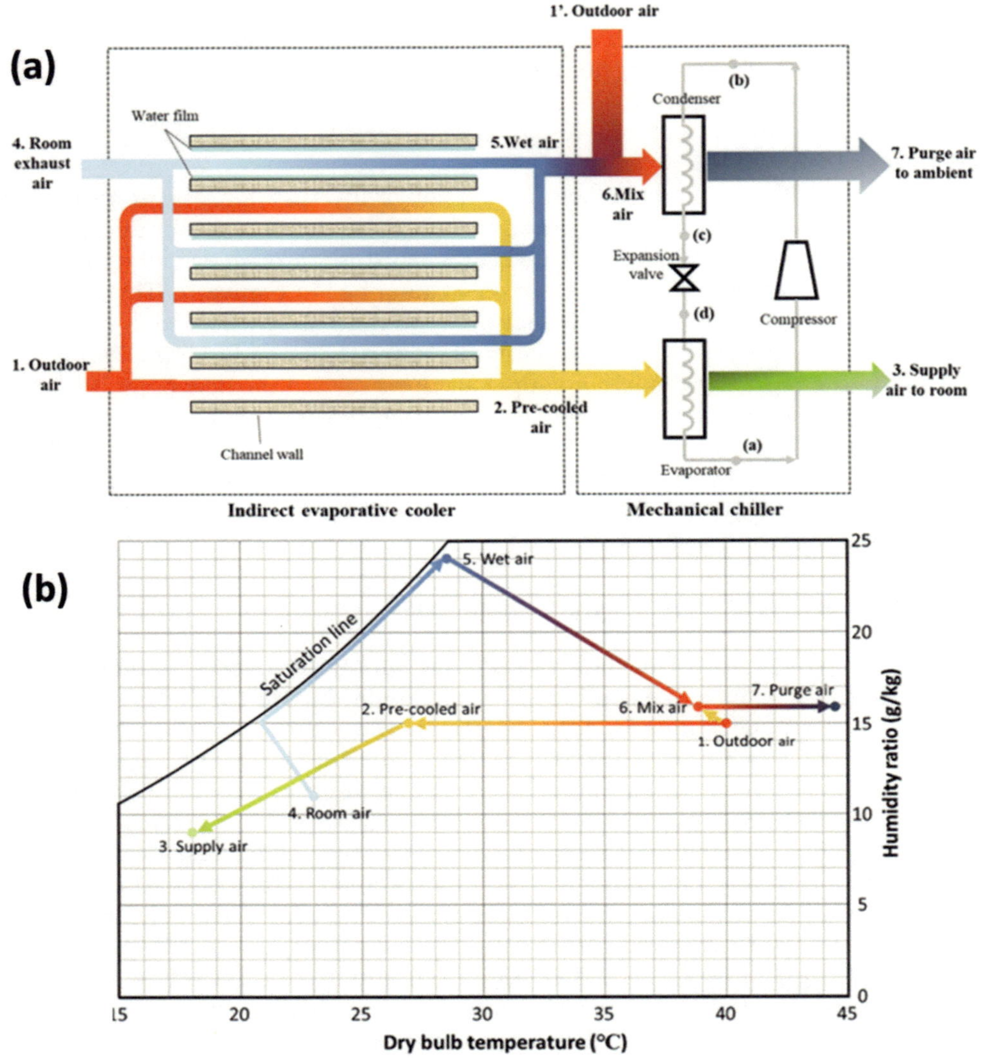

Figure 2.
(a) Schematic and (b) psychrometric representation of a hybrid IEC-MVC system obtained from [43].

3. Case studies of indirect evaporative cooling systems

3.1 Humans' thermal comfort

3.1.1 Residential buildings

Human thermal comfort is the basic requirement, which mainly depends on the operative temperature, relative humidity, and air velocity of the indoor space. The *American Society of Heating, Refrigerating and Air-Conditioning Engineers* (ASHRAE, Standard 55) guidelines define the human thermal comfort zones as 22.2 to 26.7°C for summer and 20 to 23.3°C for winter, with relative humidity levels between 30% and

Attributes	IEC	Advanced IEC	Hybrid IEC–MVC
Cooling performance	• Up to dry bulb temperature • climate-dependent	• Improved to wet-bulb temperature • Better heat and water management	• High and stable in all climates
Energy use	• Very low	• Low	• Moderate • Less than MVC)
Water use	• Moderate to high	• Optimized	• High (IEC stage)
System complexity	• Simple	• Moderate	• High
Maintenance	• Low	• Moderate	• High
Environmental impact	• Very low	• Very low	• Lower than MVC • Still use refrigerant
Applicability	• Dry climates	• Broader applications	• Versatile • High cooling loads
Cost	• Low	• Low to medium	• High

Table 1.
Summary of advantages and limitations of IEC, advanced IEC, and IEC-MVCS systems.

60% in indoor environments [46]. Numerous efforts are being made to improve human thermal comfort for enhancing performance and minimizing energy requirements. For instance, the IEC system employed for human thermal comfort achieved wet bulb effectiveness (ε_{wb}) of approximately 30 to 50% and a temperature drop of up to 5°C. It was concluded that IEC effectiveness is greatly dependent on ambient conditions and is ineffective during the hot and humid season [11]. In another study, the temperature gradient, $\varepsilon_{wb,}$ and COP were amplified to 22°C, 126–143%, and 7.15–14.58%, respectively, with improvements in IEC systems, but these were restricted to hot and dry climatic conditions. Likewise, IEC is further improved and achieves high energy efficiency, low cost, low water consumption, and a temperature gradient of up to 25°C for the hot and dry season [47]. Thus, the dependency on ambient conditions and performance degradation leads to the adaptation of the IEC-MVC systems.

Numerous studies have explored the energy efficiency and performance of IEC-MVC systems by comparing them with IEC systems. **Table 2** summarizes various studies aimed at achieving human thermal comfort under varying conditions across the globe, such as in Pakistan, Northern Italy, China, Iran, India, Singapore, and Saudi Arabia. Chen et al. [43] stated that the COP of the IEC-MVC system is 39% to 135% higher than that of the MVC system with a relatively low humidity ratio of 10 g/kg. According to **Figure 3**, the IEC system predominantly contributed to the overall COP of the hybrid IEC-MVC system, which increased with a rise in ambient temperature. Conversely, the COP of standalone MVC showed a negative relationship with increasing temperature. Similarly, the COP of IEC-MVC is higher under different temperatures (30°C–55°C) and relative humidity levels (40%–80%), and it increased with relative humidity at a constant temperature. The IEC system significantly contributes to the overall COP by reducing the energy consumption of the system [48]. It is described that IEC hybridization with MVC could save 38.2% of

Study type and location	Application/ Building type	Key findings and performance metrics	Reference(s)
Pilot experimental setup (Lab conditions)	cross-flow IEC-MVC	The hybrid IEC-MVC system reduced electricity consumption by 19% to 135% and improved COP (4.96–6.05) compared to the standalone MVC.	[43]
Dynamic simulation (Multan, Pakistan)	Building AC system (DesignBuilder-EnergyPlus)	The IEC-DEC-VCAC system provided good thermal comfort. In comparison to the standalone MVC, results showed a maximum temperature drop (21°C), increased cooling capacity, and reduced CO_2 emissions.	[45]
Dynamic simulation (21 Indian cities, five zones)	Residential building (EnergyPlus model)	IEC-DEC-DX was found to be ideal in all zones, with an improved cooling capacity and COP (up to 7), and a 25% reduction in energy consumption compared to conventional systems.	[52]
Dynamic evaluation (Saudi Arabia)	Residential/commercial cooling (arid and humid climates)	Hybrid IEC-MVC systems saved energy consumption by up to 40% in arid climates and 15–25% in humid climates.	[53]
Dynamic simulation (Beijing, China)	General building (EnergyPlus model)	The hybrid IEC-MVC system saved 38.2% more energy compared to the standalone MVC system.	[49]
Experimental (four cities in Iran)	Mechanical cooling systems	A reduction in cooling load of 75% and electricity consumption of 55% was achieved by integrating the IEC system for pre-cooling.	[50]
Experimental evaluation (Northern Italy)	Office building	Use of commercial M-cycle IEC led to a 38% reduction in total electricity consumption.	[51]
Experimental and analytical (Singapore)	Hot and humid climate	The hybrid IEC-MVC setup reduced the MVC cooling load by 32%; the model was validated under real conditions.	[54, 55]
Numerical simulation for studying the parameters affecting the IEC system performance	IEC as MVC pre-cooler (system-level)	Channel gap and cooler height were key parameters influencing cooling effectiveness.	[56]
Dynamic simulation-based study for global climates	Residential cooling (model-based)	More than 20% energy can be saved by a hybrid system in hot and arid regions, whereas modeling results were not satisfying for humid regions.	[48]

Table 2.
Summary of the hybrid IEC-MVC systems across various climates for human thermal comfort.

seasonal energy in Beijing (China) [49]; a 55% energy reduction was mentioned by Delfani et al. [50], and an experimental investigation showed a 38% reduction in total electricity requirements [51].

Despite the advantages, several limitations and factors require proper consideration for the use of hybrid IEC-MVC systems, such as water availability in water-stressed areas and maintenance for fouling, scaling, and biofilm formation.

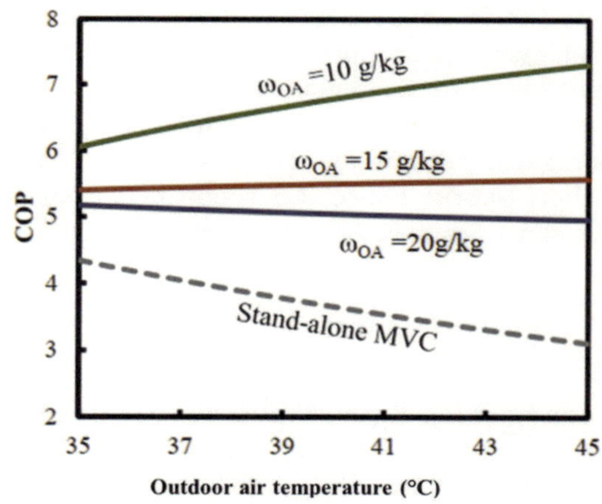

Figure 3.
Profile of COP of IEC-MVS and standalone MVC at varying outdoor temperature and humidity ratio. Solid lines represent the IEC-MVS system at 10 g/kg, 20 g/kg, and 30 g/kg of outdoor humidity ratio (ω) and the dashed line represents the COP of the standalone MVC system [43].

Also, the effectiveness of the hybrid system is based on the pre-cooling IEC system, which can be influenced by ambient conditions. In addition, the bulkiness of the hybrid system, its complex design and geometry, and the high system cost are limitations in the adaptation and commercialization of these systems.

3.1.2 Hospitals

Hospitals need to precisely regulate humidity and temperature to ensure patient thermal comfort, control airborne diseases, and prevent pathogen spread. ASHRAE Standard 170 recommends that important areas such as operating rooms and burn units maintain indoor temperatures between 28°C and 33°C and relative humidity between 30% and 60% to support skin regeneration [57]. Such strict conditions are essential for controlling infections, especially since high humidity supports bacterial survival and regrowth. In contrast, severe dryness can deactivate some viruses but damage mucosal defenses. In addition, thermal comfort and indoor air quality are critically important for patient safety. A study involving 37 hospitals in Taiwan revealed elevated CO_2 levels in wards and pharmacies due to unsatisfactory ventilation. It was found that other airborne pollutants, such as particles and microorganisms, vary with ventilation type, highlighting the necessity for efficient air management in healthcare environments [12].

Despite the IEC and IEC-MVC systems being a promising alternative for energy-efficient cooling with improved air quality, they still need to be standardized for hospitals. It is indicated that bed-integrated ventilation systems have the ability to improve energy efficiency by 83.6% and control infection risk [58]. Hence, they facilitate the use of fresh and non-recirculated air, which is a fundamental aspect of IEC technologies. Standalone IEC systems may not provide the strict humidity and temperature conditions; however, hybrid IEC-MVC systems could represent a viable option for providing energy efficiency and enhanced indoor air quality while maintaining patient comfort. An experimental and numerical study employed a series of indirect evaporative chillers in hospitals to save energy consumption and reduce pollution emissions. It is concluded that the performance favored most of the cooling season and achieved ε_{wb} of more than 106% [59]. Furthermore, some large office buildings and hospitals in dry regions are utilizing hybrid systems that combine MVC with evaporative cooling (EC), which have shown significant energy savings [60]. In this regard, Pakistan and other underdeveloped countries have limited burn units, where hybrid IEC-MVC systems may provide an energy-efficient solution to satisfy the strict temperature and humidity requirements of future burn care facilities.

3.1.3 Superstores

Supermarket air conditioning requires intensive energy for precise temperature and humidity control environments to preserve both product integrity and thermal comfort. According to ASHRAE standard guidelines, the temperature and relative humidity of indoor spaces need to be maintained in the range of 23–25°C and 45–55%, respectively, to preserve edible products and to avoid condensation on storage shelves [61]. It is stated that 17% of annual cooling energy is reduced by strategizing night ventilation as a low-energy approach with the HVAC system [62]. Likewise, sub-coolers were integrated with conventional refrigeration systems to manage energy for improving COP and subsequently enhance energy efficiency [18]. In another study, vapor injection was coupled with a compression system and achieved a COP of up to 32% compared to a standalone compression system [63]. Despite these systems, the hybridization with IEC systems offers a more sustainable strategy to improve the required conditions along with enhanced energy savings. For example, a study utilized evaporative coolers in combination with refrigeration systems. Results show an increment of up to 35% in annual energy savings compared to the refrigeration system, as shown in **Figure 4**.

Moreover, enhanced EC strategies can be suggested for supermarket applications, as pre-dehumidified air increases the system efficiency and maintains the low dew

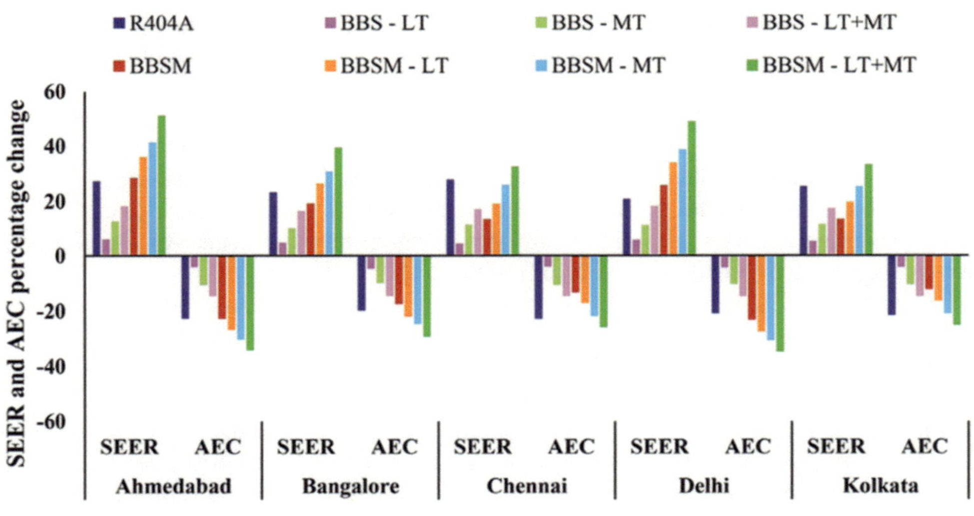

Figure 4.
Comparative study of the seasonal energy efficiency ratio (SEER) and averaged energy consumption (AEC) under various systems in comparison with a standalone refrigeration system (R404A) [65].

point. ASHRAE also highlights that removing latent heat with a standalone refrigeration system is inefficient, especially under high humidity conditions. Generally, dehumidification is achieved by hybridizing conventional mechanical dehumidification and energy-efficient desiccant-based dehumidification with post-cooling systems to control the required environments. Desiccant-based dehumidification systems are an effective approach to achieving the required humidity conditions for supermarkets. For instance, the integration of a desiccant-based hybrid system improves energy efficiency by 26–63% compared to a standalone conventional system [64].

3.1.4 Schools and academic buildings

Controlled and quality environmental conditions need to be maintained for well-being and learning performance. According to ASHRAE standards, academic buildings should maintain a temperature in the range of 22.5°C to 25.5°C, and relative humidity should not exceed 65% to ensure proper ventilation [46, 66]. Keeping in mind the air quality, indoor CO_2 levels should not exceed 700 ppm above outdoor levels (typically 300–500 ppm). A study reported that the levels of thermal comfort and CO_2 in classrooms often exceed the threshold of ASHRAE limits, which signifies the importance of improved air-conditioning design in school and academic buildings [67]. In this context, several studies highlight the potential of IEC technologies and hybrid systems for school buildings. For example, advanced IEC systems are coupled with desiccant-based air-conditioning. Results show that the employed system efficiently acquires data during the whole season with a COP of 0.43 [68]. Another simulation-based study explored the performance of the IEC-MVC system (System 1) and the IEC-MVC with improved heat and mass transfer design (System 2). It shows that System 1 achieved more than 25% energy savings in a dry climate. In comparison to System 1, System 2 reduced the energy consumption by 1 to 5 times in humid climates due to superior humidity control [69]. Kim et al. [11] evaluated indirect and direct EC systems in a campus building. The two-stage EC system achieved 51% energy savings compared to a conventional MVC system during the intermediate season, as shown in **Figure 5**. However, energy consumption increased in hot and humid climates.

3.1.5 Kitchen

In different types of residential and commercial kitchens, thermal comfort is usually influenced by increased temperatures and pollutant concentrations due to cooking activities and is even more compromised in tropical climates. Kitchen areas often surpass the upper limits of optimal temperature (23–26°C), relative humidity (40–60%), and airborne pollutants as suggested by ASHRAE Standard 55. This leads to discomfort among kitchen workers [46, 70]. A recent study performed experiments and simulations and mentioned that the supply direction and exhaust volume of air can significantly affect thermal perception and health concerns [71]. The use of air-conditioning systems with fans or modern approaches such as EC or hybrid IEC-MVC systems has shown promising results in improving thermal comfort. This system demonstrated a reduction in energy consumption of up to 17.7% with an outlet airflow rate of 5 m^3/min at 21°C [72].

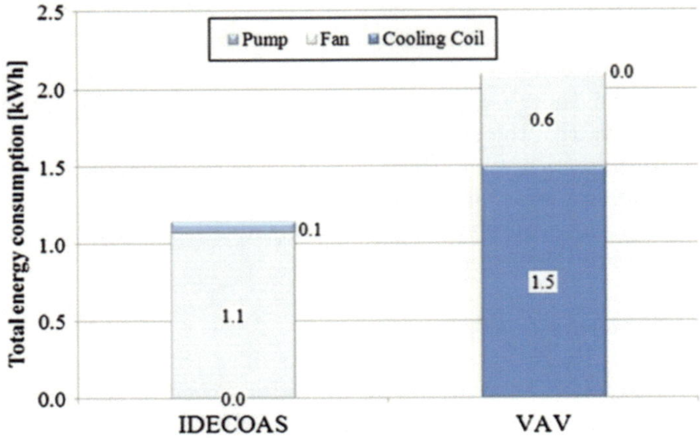

Figure 5.
Comparison of energy consumption between two-stage evaporative cooling (IDECOAS) and conventional variable air volume (VAV) systems [11].

3.1.6 Mosques

Mosques in Islamic societies serve as exclusive public spaces for religious, social, and educational activities. As a result, mosques require special thermal comfort strategies. In such environments, temperatures between 23 and 26°C, with a relative humidity below 60%, ensure acceptable human thermal comfort levels as per ASHRAE guidelines [73]. Natural ventilation and conventional HVAC systems frequently underperform during peak summer conditions in various tropical and arid regions, including the Middle East and Pakistan [74]. Numerous studies have investigated passive design and architectural modifications aimed at reducing energy consumption; however, these methods frequently face limitations due to extreme ambient conditions [75]. A study applied a passive EC strategy as a sustainable approach to achieve human thermal comfort in two different mosques in Saudi Arabia [76]. In addition, hybrid EC systems can be employed in mosques and other religious structures for consistent thermal performance and energy savings under dry climates. For example, a study used the hybrid IEC-MVC system over an extended period of time and found promising results in terms of both performance and energy savings [53]. There is a growing need to explore such climate-resilient, energy-efficient technologies through both experimental research and simulation-based studies to support sustainable cooling in energy-intensive public buildings.

3.2 Agricultural applications

These systems also offer a sustainable solution for mitigating heat stress in agriculture by maintaining optimal temperature and humidity levels. They enhance productivity in greenhouses, livestock shelters, crop storage, and so on. Key agricultural applications are discussed below.

3.2.1 Greenhouse dehumidification

The regulation of temperature and humidity in greenhouses is vital to ensure optimal photosynthesis, transpiration, and plant health. Thus, maintaining microclimatic and vapor pressure deficit (VPD) conditions becomes an essential aspect of greenhouse operations [8, 77]. Different crops need specific temperature and relative humidity preferences for optimal growth, as described in **Table 3**. For instance, tomatoes require day temperatures of 22–28°C and night temperatures of 17–20°C, with relative humidity levels maintained between 60% and 70%. To maintain such microclimatic conditions, conventional ventilation and HVAC systems often fall short under fluctuating or extreme climates, particularly in hot-humid regions [78]. Besides, these approaches are energy-intensive, promote a carbon footprint by using harmful refrigerants, and struggle to maintain optimal humidity conditions efficiently [79, 80]. Although the IEC systems have been effectively used in arid climates to maintain stable indoor conditions [34], their inability to handle latent heat loads in high-humidity conditions makes them less suitable as a standalone solution in greenhouse applications [81].

The integration of a dehumidification system is an attractive solution to address such restrictions in modern greenhouse design, as IEC systems can be coupled with desiccant dehumidification [82], MVC systems [83], or solar-assisted technologies [84]. A study explored solar-based desiccant and EC systems and achieved a seasonal COP of 2, with satisfying performance under warm ambient conditions [84]. Likewise, another study achieved an exceptionally high COP of >20 in a hybrid solid desiccant-based IEC system [85]. A study by Gadallah et al. proposed a two-stage desiccant system integrated with the advanced IEC system, achieving a COP of 1.77 during peak cooling hours, with an average COP of 0.46 due to high heat input demands [86]. The hybrid IEC-MVC system exhibits a coefficient of performance (COP) ranging from 4.96 to 6.05, resulting in a reduction of electricity consumption by 19% to 135% when compared to a standalone MVC system [43]. The studies emphasize the capabilities of hybrid and integrated systems in achieving energy-efficient climate control within agricultural greenhouses.

Plant type	T_{day} (°C)	T_{night} (°C)	RH (%)
Tomato	23–27	13–16	50–80
Cucumber	25–30	16–18	70–90
Cabbage	15–16	2	70–80
Peas	25–30	16–18	70–80
Pepper	22–30	14–16	50–70
Beans	22–26	16–18	70–80
Aubergine	25–28	14–16	50–60
Lettuce	24–28	13–16	6080
Courgettes	20–22	17–18	65–80
Strawberry	20–26	13–16	50–65

Table 3.
Requirements for temperature and relative humidity levels of different plants in a greenhouse [78].

3.2.2 Fruits and vegetables storage

Proper temperature and humidity control are critical for the post-harvest storage of fruits and vegetables to maintain freshness, extend shelf life, and reduce spoilage. For instance, leafy greens typically need temperatures close to 0°C to 2°C, and fruits like apples (7–10°C), guavas (5–10°C), mangoes (13°C), and pomegranates (0–5°C) require slightly warmer conditions with high relative humidity (90–95%) [25]. Fresh dates necessitate the MVC system [8], which is efficient in sustaining low temperatures; however, it is energy-intensive and frequently lacks flexibility in humidity control.

In this context, energy-efficient alternatives like passive EC, IEC, and hybrid IEC-MVC systems offer sustainable solutions. For instance, a study evaluated the potential of passive EC to preserve fresh fruits and vegetables longer after harvest under different regional climatic conditions [87]. An integrated IEC and EC system with solar power was used for storing tomatoes under hot, humid conditions. This system helps in reducing storage temperatures from 7 to 16°C and increasing relative humidity from 13 to 41% compared to ambient conditions, achieving cooling efficiency of up to 95.6% [88]. IEC systems can efficiently precool storage air without elevating humidity, rendering them appropriate for arid areas. Integration with MVC can provide accurate regulation of temperature and humidity. A hybrid cold storage system decreased energy consumption by up to 17.9% while maintaining an optimal environment for bananas. This approach has the benefit of managing post-harvest losses along with low-carbon emissions [89]. Therefore, the reduced energy requirements and carbon emissions of these systems are significant advantages over conventional methods, which are also suitable options for rural or off-grid agricultural practices.

3.2.3 Seeds and dried fruit storage

For seeds and dried fruit storage, cool, dry, and stable environmental conditions are required to maintain product quality, avoid fungal attacks, and enhance shelf life. As per ASHRAE guidelines in the Handbook – HVAC Applications, temperatures between 4°C and 18°C with relative humidity below 50% (30–40% for longer periods) are suitable conditions for seed storage. Dried fruits, being hygroscopic, are similarly sensitive to moisture and are best stored at temperatures below 20°C and relative humidity (RH) levels between 40% and 50%. It has been shown that intermediate RH levels during storage are critical for maintaining the physical stability of dried figs, emphasizing the need for precise humidity control [90]. Temperatures between 4°C and 18°C with relative humidity below 50% (30–40% for longer periods) are suitable conditions for seed storage. Dried fruits, being hygroscopic, are similarly sensitive to moisture and are best stored at temperatures below 20°C and RH levels between 40% and 50%. It has been shown that intermediate RH levels during storage are critical for maintaining the physical stability of dried figs, emphasizing the need for precise humidity control [26]. In places without access to the grid or in remote areas where traditional power sources are not viable, these hybrid systems provide an affordable, energy-efficient, and climate-resilient alternative.

3.2.4 Dairy and milk product storage

Dairy and milk products are highly sensitive to temperature and humidity conditions, requiring adequate storage environments to maintain their quality and safety. According to ASHRAE [91], milk should be stored at 1–4°C with 85–95% relative humidity to control microbial growth and dehydration of products like cheese and curd. It is observed that the quality of whole milk powder is maintained when stored at - 20°C for 40 days. However, both freezing and elevated temperatures degrade the quality. Similarly, the quality of fermented milk and cheeses also degrades when the storage temperature increases from 8°C (optimal storage temperature) to 14°C [92]. The EC systems provide a low-cost, practical alternative for rural and remote locations with limited refrigeration facilities. For instance, about a 10.5°C temperature drop was achieved using a charcoal-based evaporative cooler in an arid region, which extended the storage period to over 10 hours with acceptable biological quality compared to ambient storage [93]. Building on this, milk quality is more acceptable compared to fresh and chilled samples when stored using a combination of thermization (65°C) with EC [27]. Additionally, the shelf life of raw milk was increased up to 22 hours with the help of the hybrid EC system in rural and remote areas [94]. These studies collectively emphasize the scope and potential for energy-efficient and effective hybrid EC systems to enhance the shelf life of dairy products, particularly in rural and off-grid regions.

3.2.5 Livestock farming

Considering the thermal comfort of livestock, farming design requires consideration of complex environmental variables, including temperature, humidity, and ventilation rate, which directly influence livestock health and productivity [95]. ASHRAE guidelines indicate that optimal air temperature, humidity levels, and ventilation requirements vary by animal species. Cows (500 kg) require airflow ranging from 17 to 220 L/s, contingent upon the season, whereas pigs necessitate 3 to 60 L/s, depending on their growth stages [57, 95–97]. Economic constraints in numerous developing countries frequently hinder the implementation of standard air-conditioning systems, resulting in reliance on cheaper, less efficient alternatives. The precise calculation of ventilation and air supply rates is crucial, as these factors influence both sensible and latent cooling loads, thereby ensuring effective airflow and the maintenance of necessary temperature and humidity levels. Animals typically necessitate elevated humidity levels and exhibit reduced sensitivity to temperature fluctuations compared to humans; nonetheless, optimal productivity is contingent upon sustaining conditions tailored to each breed [98].

To tackle these challenges sustainably, various evaporative and hybrid cooling technologies have been examined in Pakistan and comparable climates. A study experimentally assessed the DEC, indirect IEC, and advanced IEC systems in the context of livestock applications. IEC and advanced IEC systems decreased ambient air temperature by 6.8°C and 8.9°C, respectively, creating favorable conditions for livestock, whereas DEC led to increased humidity [28]. In addition, advanced IEC-assisted desiccant air-conditioning (IEC-DAC) systems achieved a COP of 2.3 and were effective from March to October, while standalone DAC had limited performance (COP ~ 1.46), as shown in **Figure 6** [99]. Sultan et al. [98, 100] hybridized advanced IEC and solar chimney. Results showed efficient performance

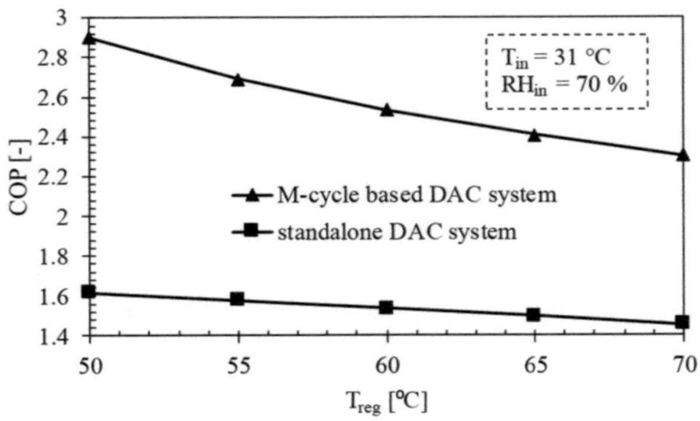

Figure 6.
Comparison of COP between IEC-assisted desiccant air-conditioning (IEC-DAC) and standalone desiccant air-conditioning (DAC) for the climatic conditions of Multan [99].

in moderate and dry climates, but performance reduced in humid climates. An EC system was investigated to alleviate heat stress in foreign dairy cows subjected to elevated temperature and humidity levels. The system significantly enhanced thermal comfort relative to energy-intensive MVC systems, providing a sustainable and cost-effective option for livestock in climate-impacted areas [101]. The hybrid IEC-MVC systems provide an energy-efficient option for regulated environments, particularly in hot and humid areas, for specific types of animals that are sensitive to temperature changes in farming or exotic animals housed in zoos. In off-grid rural areas, these investigations confirm that hybrid or solar-assisted EC systems can keep animals warm and comfortable without breaking the bank.

3.2.6 Poultry farms

Poultry birds are highly susceptible to thermal stress due to their inability to sweat, relying instead on panting to regulate body heat. According to ASHRAE guidelines, optimal conditions for broilers are a temperature range of 21 to 26°C and 50–70% relative humidity, which vary with age and variety [97]. Temperature requirements decrease from young chicks (32–35°C initially) as the birds age. Traditional MVC systems are efficient yet energy-consuming and financially unfeasible for numerous chicken farms in developing countries. In response, numerous studies have investigated energy-efficient solutions employing EC systems. The DEC, IEC, and advanced IEC systems have been examined under the climatic conditions of Multan. The innovative IEC system surpassed competitors by lowering ambient temperatures by as much as 12°C and attaining thermal comfort with a minimal temperature-humidity index (THI) [102]. Similarly, DEC and IEC systems were evaluated under various water flow rates. It was shown that DEC with $4\,L/min \cdot m^2$ and IEC with $2\,L/min \cdot m^2$ were effective, achieving significant cooling efficiencies (75–84%) and higher poultry growth with minimal energy consumption [103].

Furthermore, a desiccant dehumidification integrated IEC system with localized ventilation showed improvement in poultry farming. For instance, advanced IEC with localized air distribution improved thermal comfort and air quality in semi-arid poultry, while consuming less energy compared to conventional systems [104]. Raza et al. [41] compared the effectiveness of DEC, IEC, and advanced IEC. Results showed that the advanced IEC achieved the highest temperature drop of 10.2°C. However, performance was compromised, regardless of system type, during monsoon seasons due to high ambient humidity. A hybrid desiccant-assisted advanced IEC was proposed to improve dehumidification (4.3 g/kg–DA) and thermal comfort (THI < 30 °C) year-round, especially when polymeric desiccants were utilized [105]. These results highlight the viability of hybrid EC-desiccant systems as environmentally friendly and economically viable options for controlling the temperature of chickens in hot and humid environments.

3.2.7 Tractor cabin

Thermal comfort for tractor operators is essential in agricultural activities, particularly in harsh climatic circumstances like those in Pakistan, where the majority of tractors do not have factory-installed air-conditioned cabins. To address this, Riaz et al. [31] created and tested an air-conditioned chamber in both standard (S–I) and improved (S–II) air distribution situations. The S–II met the ISO 14269–2 comfort standards by maintaining the cabin temperature at an average of 27.4°C and the relative humidity at around 53%. Additionally, CFD models closely matched the findings of real-world tests, which confirmed expectations about airflow and comfort. This study complements past research demonstrating the intricacies of vehicle microclimates and operator comfort variables. Subsequent research, building on prior initiatives, broadened the assessment to other processes, validating a consistent temperature reduction of 11.7–12.9 °C [106]. CFD experiments showed that high vent airspeed is important for keeping operators comfortable, even when the intake temperature is high. IEC-MVC systems are a viable, energy-efficient option in this situation. Using evaporative pre-cooling before MVC may lower the strain on the compressor and make cooling more efficient. Implementing IEC-MVC in tractor cabs might be an effective way to save energy and enhance operator comfort, especially in hot and dry conditions.

3.3 Industrial applications

These systems also offer sustainable solutions for managing thermal loads in industrial environments by maintaining optimal temperature and humidity conditions. They help improve equipment performance, product quality, and worker comfort across various sectors. Key industrial applications are discussed below.

3.3.1 Tobacco processing

For tobacco processing, temperature and humidity conditions are important in different stages, including curing, softening, and storage. These controlled environmental conditions help in maintaining leaf texture, aroma, and reducing wastage. For example, tobacco leaves are processed at 30–32°C temperature and 85–88% relative humidity to rehydrate and restore pliability during the softening phase. For storage and further processing, 60–70% relative humidity and 21–24°C temperature are suitable to avoid mold formation or over-drying. A study investigated the impact of relative humidity on cigar leaves during the browning stage. Tobacco cured at 70% relative humidity demonstrated excellent chemical composition and superior leaf coloration. The relative humidity of 60% to 80% influences moisture retention and oxidation, underscoring the stringent conditions necessary for tobacco manufacturing and storage [107].

Integrated desiccant-based dehumidification systems with varying energy capacities are extremely ideal for addressing these dynamic climatic needs at each processing step. These devices can condition air with significant moisture extraction capability while preserving thermal efficiency. During curing phases characterized by elevated internal moisture levels, regenerable desiccants effectively decrease relative humidity without excessive cooling, while in softening, humidification-enhanced loops may introduce regulated moisture. Hybrid systems in storage zones provide stable humidity and temperature levels with minimal energy use, essential for preserving product quality and minimizing spoilage.

3.3.2 Matches manufacturing

In matchstick manufacturing, precise temperature (25–30°C) and humidity (50–60%) control are essential during wood seasoning, match head drying, and packaging. Improper relative humidity can cause splint warping, chemical instability, or uneven coating of ignition materials. For storage, relative humidity must stay below 65% to prevent degradation or accidental ignition. To ensure suitable conditions, desiccant-based dehumidifiers have the ability to manage moisture content without over-cooling. Considering the drying and safe storage of tobacco, an innovative IEC system can be employed for high energy efficiency and cost-effectiveness, which can further serve in humid climates and developing nations.

3.3.3 Rubber goods

During the manufacturing of rubber goods, stages such as proper mixing, molding, vulcanization, and storage require controlled temperature (22–28°C) and humidity (45–55%) conditions. Conversely, inadequate conditions can lead to defects, for example, cracking, porosity, and aging. Meanwhile, energy consumption is another concern in rubber manufacturing, especially in processes like drying and molding, which are energy-intensive. In rubber products, heaters and cooling systems are among the leading energy consumers, where applying energy-efficient technologies has yielded considerable savings [108]. Another study evaluated the most energy-intensive stages in the rubber industry and suggested an energy-saving strategy [109]. Similarly, a 60–77% energy reduction in rubber drying can be achieved by using a hybrid of solar air heating and heat recovery systems [110]. To maintain stable humidity across sensitive stages like drying and curing, desiccant-based dehumidification, especially rotary desiccant wheels, offers precise control even in low-humidity industrial environments [111]. These systems are energy-efficient, capable of extensive dehumidification, and may be run constantly. When paired with IEC or advanced IEC, the combined system provides temperature and humidity management while using substantially less energy than traditional HVAC installations. Tropical regions, with their high ambient temperatures and humidity, present unique challenges to traditional climate control systems; this hybrid technique is well-suited to these regions. In the end, using these integrated solutions helps rubber processing factories become more sustainable by improving product uniformity and reducing emissions.

Work of art material	Temperature (°C)	Relative humidity (%)
Watercolors, drawings	19–24	45–60
Documents, file materials	13–18	50–60
Ceramics, stoneware, terracotta	Not significant	Not significant
Objects from damp excavation areas	19–24	Saturated air (100%)

Table 4.
Optimal temperature and humidity parameters for various artifact types.

3.3.4 Artifacts and museums

Maintaining cultural heritage in museums requires a meticulously regulated indoor environment, where temperature and relative humidity are the most essential factors. Variations in these conditions may result in irreversible damage, including desiccation, cracking, salt crystallization, warping, and biological degradation of sensitive materials [112, 113]. ASHRAE guidelines [114] underscore the importance of stability in temperature and relative humidity with a precision level from AA to D, which is established according to associated risk levels. Level AA (risk-free) stipulates that relative humidity (RH) and temperature variations must remain within ± 5% and ± 2°C, respectively. **Table 4** presents recommended environmental parameters for different artifact categories [115].

Recent research investigates sustainable microclimatic solutions as alternatives to energy-intensive conventional HVAC systems. A CFD-based simulation research assessing HVAC setups in museum contexts revealed that desiccant-wheel systems enhanced relative humidity management during summer and decreased energy consumption by 11% [116]. In a similar vein, EC systems, including indirect-direct EC combined with humidification, have demonstrated efficacy in the preservation of unearthed relics [117]. The results indicated that an integrated IEC system with humidification could maintain relative humidity at 100% and temperature at 24.4°C with low variation, meeting ASHRAE Level AA standards, while restoring moisture to parched soil and decreasing energy usage by 40% [118]. Likewise, the integration of IEC with vapor compression cooling (IEC-MVC) improves efficacy in the presence of variable ambient humidity [119]. This approach can be implemented in accordance with the necessary and ambient environmental conditions. Future museum HVAC designs can greatly benefit from such energy-efficient IEC and/or hybrid systems that balance microclimatic control, artifact preservation, and energy efficiency.

3.3.5 Mobile storage/marine

Due to the growing requirement for diverse postharvest preservation solutions in transportation, offshore operations, and maritime logistics, efficient and portable cold storage systems have surged. EC systems' energy economy and simple construction make them suitable for mobile usage. They're not ideal for moist or confined spaces unless you adjust them for humidity and temperature.

Hybrid systems that integrate the IEC with desiccant-based dehumidification and additional modifications have demonstrated considerable potential in addressing these limitations. A commercial-scale hybrid IEC utilizing a latent heat thermal energy storage system with phase change materials significantly improves air pre-cooling and humidity reduction, thereby enhancing energy efficiency in hot climates [120]. In land mobile units, an evaporative membrane cooling system for motorhomes and vehicles indicates mechanical energy savings of up to 36–38% under moderate humidity levels. This was achieved by pre-cooling the incoming air to lower the load of refrigerant vapor while maintaining the efficiency of the airflow [121]. In addition, enhanced mobile air-conditioning using compact microchannel heat exchangers resulted in a 5–8% improvement in cooling capacity and COP for vehicle cabins [122]. A mobile IEC was integrated with an EC system featuring a 3.8-ton capacity for tomato storage, effectively maintaining optimal relative humidity and temperature in sub-humid conditions while facilitating transport from field to market. Additional enhancements, such as automation and operation using vehicle kinetic energy, were suggested in the research [88].

In the maritime sector, energy consumption is also a concern. Although a passive approach, like refrigerated container arrangement, may reduce energy consumption to some extent, it is not consistent for all containers [123]. Furthermore, Chen et al. [124] applied regenerative IEC systems integrated with MVC in wet markets, repurposing condensate water for cooling and yielding substantial energy savings. A promising marine-specific innovation is the seawater-based IEC integrated with desiccant wheel systems that is modeled. This hybrid system is tested using seawater as the evaporative medium and residual ship heat to drive regeneration. Results showed a high COP value of 12.6 with up to a 61.6% reduction in energy, as illustrated in **Figure 7**. It is proven to be a practical, sustainable air-conditioning solution in ship cabins [19]. These developments emphasize the adaptability of hybrid IEC-desiccant systems for mobile and marine settings. Their capacity to provide superior energy efficiency while ensuring accurate temperature and humidity regulation in fluctuating situations renders them ideal for cold chain logistics, rural applications, and maritime operations.

To summarize, hybrid EC systems provide a sustainable and energy-efficient alternative for maintaining thermal comfort in residential, commercial, and

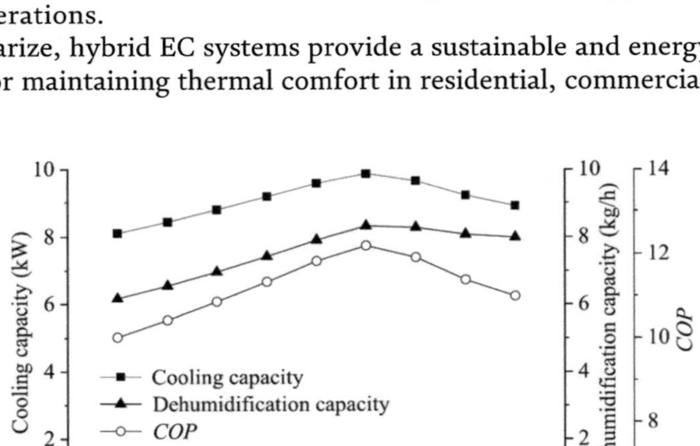

Figure 7.
Variations in the influence of COP, dehumidification capacity, and cooling capacity with changing outdoor air temperature [19].

industrial buildings, as well as agricultural and industrial applications. Their capacity to manage temperature and humidity with minimal environmental impact makes them a viable choice for all applications requiring dynamic climatic and operating conditions.

4. Conclusion

Hybrid indirect EC systems have emerged as a promising sustainable solution for providing thermal comfort and humidity control across a wide range of climates and applications. Their adaptability allows for integration into residential, commercial, agricultural, and industrial settings, delivering efficient cooling with significantly reduced energy consumption compared to conventional systems. Case studies from various sectors show that the integration of IEC can maintain indoor thermal environments within optimal comfort or process-specific ranges, even under challenging ambient conditions.

- Hybrid IEC systems significantly reduce cooling energy consumption compared to conventional vapor-compression systems, particularly in hot-dry and hot-humid climates.

- These systems are effective in maintaining thermal comfort across residential, commercial, agricultural, and industrial sectors, including greenhouses, cold storage facilities, poultry houses, museums, and vehicle cabins.

- Integrating IEC systems with MVC, desiccant dehumidification, solar energy, and so on improves humidity control, operational flexibility, and overall cooling performance.

- Hybrid IEC systems are adaptable to variable climatic conditions, offering viable solutions even in humid environments where traditional evaporative systems are less effective.

- Portable and marine applications benefit from compact IEC designs with energy recovery options, enabling flexible deployment for cold-chain and rural scenarios.

- These systems minimize greenhouse gas emissions and dependency on toxic refrigerants, aligning with low-carbon and sustainable development targets.

- For large-scale adoption, challenges such as system complexity, water availability, and initial cost must be addressed through policy support, local manufacturing, and user training.

Future work should focus on developing region-specific design guidelines, improving component efficiency, and integrating renewable energy sources to further enhance system sustainability and accessibility. Public and private sector collaboration can accelerate adoption through demonstration projects, subsidies, and standardization efforts.

4.1 CRediT authorship contribution statement

Muhammad Ali Imran: Conceptualization, methodology, software, formal analysis, investigation, and writing – original draft. **Muhammad Sultan, Muhammad Wakil Shahzad:** Conceptualization, methodology, validation, resources, visualization, supervision, project administration, writing – review and editing, and funding acquisition. **Muhammad Ahmad Jamil, Uzair Sajjad, Zhaoli Zhang, Hafiz M.U. Raza, Muhammad Haseeb Siddiqui, Muhammad Asad Ali, and Muhammad Hamid Mahmood:** Conceptualization, methodology, software, formal analysis, investigation, and writing – review and editing.

Funding

The UK Government funds this research and/or pilot activity through Project: APP47457, titled "Super-efficient Sustainable Cooling Solution for All Applications (S2Cool)" under the Ayrton Challenge Programme of UK Research and Innovation (UKRI); however, the views expressed do not necessarily reflect the UK Government's official policies. Project: APP47457 is implemented and led by Northumbria University, UK, together with partners to develop sustainable cooling solutions for several applications.

Data availability statement

The datasets used and/or analyzed during the current study are available from the corresponding authors upon request.

Acknowledgment

The UK Government funds this research and/or pilot activity through the project APP47457, titled "Super-efficient Sustainable Cooling Solution for All Applications (S2Cool)," under the Ayrton Challenge Program of the UK Research and Innovation (UKRI); however, the views expressed do not necessarily reflect the UK Government's official policies. The project APP47457 is implemented and led by Northumbria University, UK, together with partners to develop sustainable cooling solutions for several applications.

Conflict of Interest

The authors declare no conflict of interest.

Author details

Muhammad Ali Imran[1], Muhammad Sultan[1*], Muhammad Wakil Shahzad[2*], Muhammad Ahmad Jamil[2], Hafiz M.U. Raza[1], Muhammad Hamid Mahmood[1], Muhammad Asad Ali[1], Muhammad Haseeb Siddiqui[1], Uzair Sajjad[3] and Zhaoli Zhang[4]

1 Department of Agricultural Engineering, Bahauddin Zakariya University, Multan, Pakistan

2 Department of Mechanical and Construction Engineering, Northumbria University, Newcastle upon Tyne, United Kingdom

3 Interdisciplinary Research Center for Sustainable Energy Systems (IRC-SES), King Fahd University of Petroleum and Minerals, Dhahran, Saudi Arabia

4 School of Mechanical Engineering, Southwest Jiaotong University, Chengdu, China

*Address all correspondence to: muhammadsultan@bzu.edu.pk

References

[1] Ahmad T, Zhang D. A critical review of comparative global historical energy consumption and future demand: The story told so far. Energy Reports. 2020;**6**:1973–1991

[2] Ghasemian S, Faridzad A, Abbaszadeh P, Taklif A, Ghasemi A, Hafezi R. An overview of global energy scenarios by 2040: Identifying the driving forces using cross-impact analysis method. International Journal of Environmental Science and Technology. 2024;**21**:7749–7772

[3] Andersen SO, Halberstadt ML, Borgford-Parnell N. Stratospheric ozone, global warming, and the principle of unintended consequences—An ongoing science and policy success story. Journal of the Air & Waste Management Association. 2013;**63**:607–647

[4] Zhao N, Elshareef H, Li B, Wang B, Jia Z, Zhou L, Liu Y, Sultan M, Dong R, Zhou Y. The efforts of China to combat air pollution during the period of 2015–2018: A case study assessing the environmental, health and economic benefits in the Beijing-Tianjin-Hebei and surrounding "2+26" regions. Science of the Total Environment. 2022;**853**:158437

[5] Asfahan HM, Sajjad U, Sultan M, Hussain I, Hamid K, Ali M, Wang -C-C, Shamshiri RR, Khan MU. Artificial intelligence for the prediction of the thermal performance of evaporative cooling systems. Energies. 2021;**14**:3946. DOI: 10.3390/en14133946

[6] Kapilan N, Isloor AM, Karinka S. A comprehensive review on evaporative cooling systems. Results in Engineering. 2023;**18**:101059. DOI: 10.1016/j.rineng.2023.101059

[7] Xue T, Wan Y, Huang Z, Chen P, Lin J, Chen W, Liu H. A comprehensive review of the applications of hybrid evaporative cooling and solar energy source systems. Sustainability. 2023;**15**:16907. DOI: 10.3390/su152416907

[8] Sultan M, Miyazaki T. Energy-Efficient Air-Conditioning Systems for Nonhuman Applications. Refrigeration. 2017;**10**:97–117

[9] Wasti TZ, Sultan M, Aleem M, Sajjad U, Farooq M, Raza HM, Khan MU, Noor S. An overview of solid and liquid materials for adsorption-based atmospheric water harvesting. Advances in Mechanical Engineering. 2022;**14**:16878132221082768

[10] Ali M, Ahmad W, Sheikh NA, Ali H, Kousar R, Ur Rashid T. Performance enhancement of a cross flow dew point indirect evaporative cooler with circular finned channel geometry. Journal of Building Engineering. 2021;**35**:101980

[11] Kim M-H, Jeong J-W. Cooling performance of a 100% outdoor air system integrated with indirect and direct evaporative coolers. Energy. 2013;**52**:245–257. DOI: 10.1016/j.energy.2013.02.008

[12] Jung -C-C, Wu P-C, Tseng C-H, Su H-J. Indoor air quality varies with ventilation types and working areas in hospitals. Building and Environment. 2015;**85**:190–195

[13] Wu HT, Li QS, Dai RC, Liu S, Wu L, Mao W, Ji CH. Effects of air-conditioning systems in the public areas of hospitals: A scoping review. Epidemiology and Infection. 2021;**149**:e201

[14] Zorpas AA, Skouroupatis A. Indoor air quality evaluation of two museums in a subtropical climate conditions. Sustainable Cities and Society. 2016;**20**:52–60

[15] Verde M, Harby K, Corberán JM. Optimization of thermal design and geometrical parameters of a flat tube-fin adsorbent bed for automobile air-conditioning. Applied Thermal Engineering. 2017;**111**:489–502

[16] Lee S, Lee WL. Site verification and modeling of desiccant-based system as an alternative to conventional air-conditioning systems for wet markets. Energy. 2013;**55**:1076–1083

[17] Soon JM, Wahab IRA. On-site hygiene and biosecurity assessment: A new tool to assess live bird stalls in wet markets. Food Control. 2021;**127**:108108

[18] Yang L, Zhang C-L. Analysis on energy saving potential of integrated supermarket HVAC and refrigeration systems using multiple subcoolers. Energy and Buildings. 2010;**42**:251–258

[19] Liu L, Chen Y, Yan H, Peng S, Pan Y. Study on performance of seawater-based evaporative cooling integrated with desiccant wheel air-conditioning system for marine vessels. Applied Energy. 2025;**389**:125723

[20] Kowalski P, Kwiecień D. Evaluation of simple evaporative cooling systems in an industrial building in Poland. Journal of Building Engineering. 2020;**32**:101555

[21] Naşcu I, Diangelakis NA, Muñoz SG, Pistikopoulos EN. Advanced model predictive control strategies for evaporation processes in the pharmaceutical industries. Computers and Chemical Engineering. 2023;**173**:108212

[22] Saghafifar M, Gadalla M. Innovative inlet air cooling technology for gas turbine power plants using integrated solid desiccant and Maisotsenko cooler. Energy. 2015;**87**:663–677

[23] De Antonellis S, Joppolo CM, Liberati P, Milani S, Romano F. Modeling and experimental study of an indirect evaporative cooler. Energy and Buildings. 2017;**142**:147–157. DOI: 10.1016/j.enbuild.2017.02.057

[24] Mahmood MH, Sultan M, Miyazaki T. Significance of temperature and humidity control for agricultural products storage: Overview of conventional and advanced options. International Journal of Food Engineering. 2019;**15**. DOI: 10.1515/ijfe-2019-0063

[25] Lal Basediya A, Samuel DV, Beera V. Evaporative cooling system for storage of fruits and vegetables - a review. Journal of Food Science and Technology. 2013;**50**:429–442. DOI: 10.1007/s13197-011-0311-6

[26] Mahmood MH, Sultan M, Miyazaki T. Experimental evaluation of desiccant dehumidification and air-conditioning system for energy-efficient storage of dried fruits. Building Services Engineering Research and Technology. 2020;**41**:454–465

[27] Sempiira EJ, Mugisa DJ, Galiwango J, Kisaalita WS. Combining thermization and evaporative cooling toward milk freshness preservation at the smallholder farm level. Journal of Food Process Engineering. 2020;**43**: e13529

[28] Raza HMU, Sultan M, Bahrami M, Khan AA. Experimental investigation of evaporative cooling systems for agricultural storage and livestock air-conditioning in Pakistan. Building Simulation. 2020;**14**:617–631. DOI: 10.1007/s12273-020-0678-2

[29] Riaz N, Sultan M, Ashraf H, Mahmood MH, Ahmad F, Khan MU, Farooq M, Sajjad U, Shahzad MW, Zhang Z. Study on evaporative cooling assisted temperature and humidity control systems for greenhouse farming in Pakistan. In: Sultan M, editor. Air Conditioning Technologies and Applications. London, UK: Intechopen; 2024. DOI: 10.5772/intechopen.1003410

[30] Patel DP, Jain SK, Lakhawat SS, Wadhawan N. A low-cost storage for horticulture commodities for enhancing farmer's income: An overview on evaporative cooling. Journal of Food Process Engineering. 2022;**45**:e14134

[31] Riaz M, Mahmood MH, Ashraf MN, Sultan M, Sajjad U, Hamid K, Farooq M, Wang F. Experiments and CFD simulation of an air-conditioned tractor cabin for thermal comfort of tractor operators in Pakistan. Heliyon. 2023;**9**:e23038

[32] Ružić D, Homa D, Simikić M, Vejnović S. Comparative experimental study of microclimate conditions in farm tractor cabs. Sādhanā. 2023;**48**:258

[33] Liberty J, Ugwuishiwu B, Pukuma S, Odo C. Principles and application of evaporative cooling systems for fruits and vegetables preservation. International Journal of Current Engineering and Technology. 2013;**3**:1000–1006

[34] Duan Z, Zhan C, Zhang X, Mustafa M, Zhao X, Alimohammadisagvand B, Hasan A. Indirect evaporative cooling: Past, present and future potentials. Renewable and Sustainable Energy Reviews. 2012;**16**:6823–6850. DOI: 10.1016/j.rser.2012.07.007

[35] Hussain I, Bibi F, Bhat SA, Sajjad U, Sultan M, Ali HM, Azam MW, Kaushal SK, Hussain S, Yan W-M. Evaluating the parameters affecting the direct and indirect evaporative cooling systems. Engineering Analysis with Boundary Elements. 2022;**145**:211–223. DOI: 10. 1016/j.enganabound.2022.09.016

[36] Mahmood MH, Sultan M, Miyazaki T, Koyama S, Maisotsenko VS. Overview of the Maisotsenko cycle – A way towards dew point evaporative cooling. Renewable and Sustainable Energy Reviews. 2016;**66**:537–555. DOI: 10.1016/j.rser.2016.08.022

[37] Riangvilaikul B, Kumar S. An experimental study of a novel dew point evaporative cooling system. Energy and Buildings. 2010;**42**:637–644. DOI: 10.1016/j.enbuild.2009.10.034

[38] Cui Y, Zhu J, Zoras S, Liu L. Review of the recent advances in dew point evaporative cooling technology: 3E (energy, economic and environmental) assessments. Renewable and Sustainable Energy Reviews. 2021;**148**. DOI: 10.1016/j.rser.2021.111345

[39] Jradi M, Riffat S. Experimental and numerical investigation of a dew-point

cooling system for thermal comfort in buildings. Applied Energy. 2014;**132**:524–535

[40] Rogdakis ED, Koronaki IP, Tertipis DN. Experimental and computational evaluation of a Maisotsenko evaporative cooler at Greek climate. Energy and Buildings. 2014;**70**:497–506. DOI: 10.1016/j.enbuild.2013.10.013

[41] Raza HMU, Ashraf H, Shahzad K, Sultan M, Miyazaki T, Usman M, Shamshiri RR, Zhou Y, Ahmad R. Investigating applicability of evaporative cooling systems for thermal comfort of poultry birds in Pakistan. Applied Sciences. 2020;**10**:4445. DOI: 10.3390/app10134445

[42] Chen Q, Burhan M, Ja MK, Shahzad MW, Ybyraiymkul D, Zheng H, Cui X, Ng KC. Hybrid indirect evaporative cooling-mechanical vapor compression system: A mini-review. Energies. 2022;**15**. DOI: 10.3390/en15207810

[43] Chen Q, Kum Ja M, Burhan M, Akhtar FH, Shahzad MW, Ybyraiymkul D, Ng KC. A hybrid indirect evaporative cooling-mechanical vapor compression process for energy-efficient air conditioning. Energy Conversion and Management. 2021;**248**. DOI: 10.1016/j.enconman.2021.114798

[44] Khalajzadeh V, Farmahini-Farahani M, Heidarinejad G. A novel integrated system of ground heat exchanger and indirect evaporative cooler. Energy and Buildings. 2012;**49**:604–610. DOI: 10.1016/j.enbuild.2012.03.009

[45] Noor S, Ashraf H, Sultan M, Khan ZM. Evaporative cooling options for building air-conditioning: A comprehensive study for climatic conditions of Multan (Pakistan). Energies. 2020;**13**:3061. DOI: 10.3390/en13123061

[46] Standard A. Thermal environmental conditions for human occupancy. ANSI/ASHRAE, 55 **1992**, 5 p

[47] Jamil MA, Shahzad MW, Xu BB, Imran M, Ng KC, Zubair SM, Markides CN, Worek WM. Energy-efficient indirect evaporative cooler design framework: An experimental and numerical study. Energy Conversion and Management. 2023;**292**. DOI: 10.1016/j.enconman.2023.117377

[48] Jacob TA, Shah N, Park WY. Evaluation of hybrid evaporative-vapor compression air conditioners for different global climates. Energy Conversion and Management. 2021;**249**. DOI: 10.1016/j.enconman.2021.114841

[49] Duan Z, Zhao X, Liu J, Zhang Q. Dynamic simulation of a hybrid dew point evaporative cooler and vapour compression refrigerated system for a building using EnergyPlus. Journal of Building Engineering. 2019;**21**:287–301

[50] Delfani S, Esmaeelian J, Pasdarshahri H, Karami M. Energy saving potential of an indirect evaporative cooler as a pre-cooling unit for mechanical cooling systems in Iran. Energy and Buildings. 2010;**42**:2169–2176

[51] Zanchini E, Naldi C. Energy saving obtainable by applying a commercially available M-cycle evaporative cooling system to the air conditioning of an office building in North Italy. Energy. 2019;**179**:975–988

[52] Venkateswara Rao V, Datta SP. A feasibility assessment of single to multi/hybrid evaporative coolers for building air-conditioning across diverse climates in India. Applied Thermal Engineering. 2020;**168**. DOI: 10.1016/j.applthermaleng.2019.114813

[53] Chen Q, M KJ, Burhan M, Shahzad MW, Ybyraiymkul D, Oh S, Cui X, Ng KC. Long-term performance of a hybrid indirect evaporative cooling-mechanical vapor compression cycle: A case study in Saudi Arabia. Frontiers in Built Environment. 2022;**8**. DOI: 10.3389/fbuil.2022.1032961

[54] Cui X, Chua KJ, Islam MR, Ng KC. Performance evaluation of an indirect pre-cooling evaporative heat exchanger operating in hot and humid climate. Energy Conversion and Management. 2015;**102**:140–150. DOI: 10.1016/j.enconman.2015.02.025

[55] Cui X, Sun L, Zhang S, Jin L. On the study of a hybrid indirect evaporative pre-cooling system for various climates. Energies. 2019;**12**:4419. DOI: 10.3390/en12234419

[56] Chen Y, Yang H, Luo Y. Parameter sensitivity analysis and configuration optimization of indirect evaporative cooler (IEC) considering condensation. Applied Energy. 2017;**194**:440–453

[57] ASHRAE, A. Standard 170-2013 ventilation of health care facilities. Atlanta: American Society of Heating, Refrigerating and Air-Conditioning Engineers, Inc; 2013

[58] Wang Z, Bivolarova M, Melikov A. Energy-saving and IAQ control in hospital patient room by bed-integrated ventilation. In Proceedings of 14th REHVA HVAC World Congress; p. 1716–1723

[59] Jing Y, Zhao C, Xie X, Jiang Y. Optimization and performance analysis of water-mediated series indirect evaporative chillers: Experimental and simulated investigation. Energy Conversion and Management. 2022;**268**:115990. DOI: 10.1016/j.enconman.2022.115990

[60] Hunn BD, Peterson JL. Cost-effectiveness of Indirect Evaporative Cooling for Commercial Buildings in Texas. Atlanta, GA, United States: American Society of Heating, Refrigerating and Air-Conditioning Engineers . . .; 1996. 0001-2505

[61] Heating, A.S.o.; Refrigerating; Engineers, A.-C. ASHRAE Handbook: Refrigeration Systems and Applications; American Society of Heating, Refrigerating and Air Conditioning Engineers: 1986

[62] Mylona Z, Kolokotroni M, Tassou SA. Coupling night ventilative and active cooling to reduce energy use in supermarkets with high refrigeration loads. Energy and Buildings. 2018;**171**:26–39

[63] Jiang J, Yang Q, Yin T, Zheng Z, Zhao Y, Liu G, Li L. Comparative thermodynamic study of CO2 transcritical supermarket booster refrigeration system combined with vapor injection and parallel compression. International Journal of Refrigeration. 2025;**170**:70–84

[64] Lazzarin RM, Castellotti F. A new heat pump desiccant dehumidifier for supermarket application. Energy and Buildings. 2007;**39**:59–65

[65] Lata M, Gupta DK. Performance evaluation and comparative analysis of trans-critical CO2 booster refrigeration systems with modified evaporative

cooled gas cooler for supermarket application in Indian context. International Journal of Refrigeration. 2020;**120**:248–259. DOI: 10.1016/j.ijrefrig.2020.08.004

[66] Muller CO, Chamas H, Zouggari Y. ASHRAE Standard 62.1 and LEED: Using enhanced air cleaning to integrate IAQ and energy conservation

[67] Asif A, Zeeshan M, Jahanzaib M. Indoor temperature, relative humidity and CO2 levels assessment in academic buildings with different heating, ventilation and air-conditioning systems. Building and Environment. 2018;**133**:83–90. DOI: 10.1016/j.buildenv.2018.01.042

[68] Güzelel YE, Olmuş U, Büyükalaca O. Simulation of a desiccant air-conditioning system integrated with dew-point indirect evaporative cooler for a school building. Applied Thermal Engineering. 2022;**217**:119233

[69] Li X, Chang Y, Liu W, Yao W, Zhu G. Energy-saving potential of dew point indirect evaporative cooler applied in school building. Applied Thermal Engineering. 2025;**275**:126865

[70] Alam MS, Sharma M, Salve UR. Assessment of thermal comfort in a hot and humid indoor built environment of a kitchen at a university canteen. Work (Reading, Mass.). 2022;**72**:189–199

[71] Chen Z, Xin J, Liu P. Air quality and thermal comfort analysis of kitchen environment with CFD simulation and experimental calibration. Building and Environment. 2020;**172**:106691

[72] Zhang S, Shen Z, Chen Z. Optimization strategy for joint operation of kitchen air-conditioning and fan based on energy consumption and comfort. Building and Environment. 2025;**276**:112919

[73] Heating, A.S.o.; Refrigerating; Engineers, A.-C.; Institute, A.N.S. Thermal Environmental Conditions for Human Occupancy; American Society of Heating, Refrigerating and Air-Conditioning Engineers: 2004; Vol. **55**

[74] Mushtaha E, Helmy O. Impact of building forms on thermal performance and thermal comfort conditions in religious buildings in hot climates: A case study in Sharjah city. International Journal of Sustainable Energy. 2017;**36**:926–944

[75] Azmi NA, Arıcı M, Baharun A. A review on the factors influencing energy efficiency of mosque buildings. Journal of Cleaner Production. 2021;**292**:126010

[76] Al-Tassan AA, Bahobail MA. Mosques and Sustainable Traditional Technique. Saudi Arabia: King Saud University; 2006

[77] Badji A, Benseddik A, Bensaha H, Boukhelifa A, Hasrane I. Design, technology, and management of greenhouse: A review. Journal of Cleaner Production. 2022;**373**:133753

[78] Abbas F, Sultan M, Shahzad MW, Farooq M, Raza HMU, Mahmood MH, Sajjad U, Zhang Z. Comprehensive review on evaporative cooling and desiccant dehumidification technologies for agricultural greenhouses. AgriEngineering. 2025;7:222. DOI: 10.3390/agriengineering7070222

[79] Sultan M, Miyazaki T, Koyama S, Khan ZM. Performance evaluation of hydrophilic organic polymer sorbents for desiccant air-conditioning applications. Adsorption Science &

Technology. 2017;**36**:311–326. DOI: 10.1177/0263617417692338

[80] Rafique MM, Gandhidasan P, Rehman S, Al-Hadhrami LM. A review on desiccant based evaporative cooling systems. Renewable and Sustainable Energy Reviews. 2015;**45**:145–159

[81] Porumb B, Ungureşan P, Tutunaru LF, Şerban A, Bălan M. A review of indirect evaporative cooling technology. Energy Procedia. 2016;**85**:461–471. DOI: 10.1016/j.egypro.2015.12.228

[82] Ashraf H, Sultan M, Shamshiri RR, Abbas F, Farooq M, Sajjad U, Md-Tahir H, Mahmood MH, Ahmad F, Taseer YR. Dynamic evaluation of desiccant dehumidification evaporative cooling options for greenhouse air-conditioning application in multan (Pakistan). Energies. 2021;**14**:1097

[83] Chen W, Cui X, Huang ZF, Shao YL, Chua KJ. Developing an integrated solid desiccant dehumidifier and dew-point evaporative cooler for green air conditioning. Journal of Building Engineering. 2024;**96**:110404. DOI: 10.1016/j.jobe.2024.110404

[84] Comino F, González JC, Navas-Martos F, De Adana MR. Experimental energy performance assessment of a solar desiccant cooling system in Southern Europe climates. Applied Thermal Engineering. 2020;**165**:114579

[85] Goldsworthy M, White S. Optimisation of a desiccant cooling system design with indirect evaporative cooler. International Journal of Refrigeration. 2011;**34**:148–158

[86] Gadalla M, Saghafifar M. Performance assessment and transient optimization of air precooling in multi-stage solid desiccant air conditioning systems. Energy Conversion and Management. 2016;**119**:187–202

[87] Defraeye T, Shoji K, Schudel S, Onwude D, Shrivastava C. Passive evaporative coolers for postharvest storage of fruit and vegetables: Where to best deploy them and how well do they perform. Frontiers in Food Science and Technology. 2023;**3**:1100181

[88] Sibanda S, Workneh TS. Performance evaluation of an indirect air cooling system combined with evaporative cooling. Heliyon. 2020;**6**: e03286

[89] Natarajan B, Kathiresan AC, Subramanium SK. Development and performance evaluation of a hybrid portable solar cold storage system for the preservation of vegetables and fruits in remote areas. Journal of Energy Storage. 2023;**72**:108292

[90] Badii F, Farahnaky A, Behmadi H. Effect of storage relative humidity on physical stability of dried fig. Journal of Food Processing and Preservation. 2014;**38**:477–483

[91] Heating, A.S.o.; Refrigerating; Engineers, A.-C. ASHRAE Handbook; American Society of Heating, Refrigerating and Air-conditioning Engineers: 1994

[92] Memiši NR, Vesković-Moračanin SM, Škrinjar MM, Iličić MD, Ač MĐ. Storage temperature: A factor of shelf life of dairy products. Acta Periodica Technologica. 2014;**45**:55–66

[93] Wayua FO, Okoth MW, Wangoh J. Design and performance assessment of a low cost evaporative cooler for storage of camel milk in arid pastoral areas of

Kenya. International Journal of Food Engineering. 2012;**8**(1):16

[94] Ridley SC, Shalo PL. Farm application of lactoperoxidase treatment and evaporative cooling for the intermediate preservation of unprocessed milk in Kenya. Journal of Food Protection. 1990;**53**:592–598

[95] ASHRAE H. Handbook—HVAC Applications. American Society of Heating Refrigerating and Air-Conditioning Engineers **2007**

[96] ASHRAE. Handbook-Fundamentals. **2009**

[97] Edition S 2011 ASHRAE HANDBOOK HVAC Applications-gearju.com

[98] Sultan M, Miyazaki T, Niaz H, Shabir F, Ashraf S, Khan Z, Mahmood M, Raza H. Thermodynamic assessment of solar chimney based airconditioning system for agricultural and livestock applications. In Proceedings of Proceedings of the 4th International Conference on Energy and Environment and Sustainable Development (EESD-2016), 2016, Jamshoro, Pakistan; p. 1–3

[99] Niaz H, Sultan M, Khan AA, Miyazaki T, Feng Y-Q, Khan ZM, Niaz Y, Mahmood MH, Ali I, Imran MA. Study on evaporative cooling assisted desiccant air conditioning system for livestock application in Pakistan. **2019**

[100] Sultan M, Miyazaki T, Mahmood MH, Khan ZM. Solar assisted evaporative cooling based passive air-conditioning system for agricultural and livestock applications. Journal of Engineering Science and Technology. 2018;**13**:693–703

[101] Ullah HS, Ashraf H, Bilal M, Sultan M, Shabir F, Ashraf S, Mahmood MH, Niaz Y, Khan ZM. Evaporative cooling systems for thermal comfort of foreign cattle breeds: THI evaluation and system feasibility. **2020**

[102] Shahzad K, Sultan M, Bilal M, Ashraf H, Farooq M, Miyazaki T, Sajjad U, Ali I, Hussain MI. Experiments on energy-efficient evaporative cooling systems for poultry farm application in Multan (Pakistan). Sustainability. 2021;**13**:2836. DOI: 10.3390/su13052836

[103] Abdel-Rahman SG. Performance evaluation of poultry houses under different evaporative cooling systems. Zagazig Journal of Agricultural Research. 2020;**47**:999–1010

[104] Al Assaad DK, Orabi MS, Ghaddar NK, Ghali KF, Salam DA, Ouahrani D, Farran MT, Habib RR. A sustainable localised air distribution system for enhancing thermal environment and indoor air quality of poultry house for semiarid region. Biosystems Engineering. 2021;**203**: 70–92

[105] Aleem M, Sultan M, Mahmood MH, Miyazaki T. Desiccant dehumidification cooling system for poultry houses in multan (Pakistan). In Energy-Efficient Systems for Agricultural Applications. Cham: Springer; 2021. 19–42

[106] Riaz M, Mahmood MH, Ashraf MN, Sultan M, Sajjad U, Riaz F, Farooq M, Zhang Z. Experimental study of an air-conditioned tractor cabin using CFD analysis for primary and secondary tillage operations in Pakistan. Results in Engineering. 2025;**26**:104540. DOI: 10.1016/j.rineng.2025.104540

[107] Zhao S, Wu Z, Lai M, Zhao M, Lin B. Determination of optimum humidity for air-curing of cigar tobacco leaves during the browning period. Industrial Crops and Products. 2022;**183**:114939

[108] Saidur R, Mekhilef S. Energy use, energy savings and emission analysis in the Malaysian rubber producing industries. Applied Energy. 2010;**87**:2746–2758

[109] Piccioni M, Martini F, Martini C, Toro C. Evaluation of energy performance indicators and energy saving opportunities for the Italian rubber manufacturing industry. Energies. 2024;**17**:1584

[110] Tan T, Zhang H, Yang M, Baeyens J. Energy-efficient crumb rubber drying by heat recovery and solar air heating. Industrial Crops and Products. 2024;**209**:117970

[111] Su X, Geng Y, Huang L, Li S, Wang Q, Xu Z, Tian S. Review on dehumidification technology in low and extremely low humidity industrial environments. Energy. 2024;**302**:131793

[112] Pavlogeorgatos G. Environmental parameters in museums. Building and Environment. 2003;**38**:1457–1462

[113] Thomson G. The museum environment. Elsevier. 2013

[114] Handbook A HVAC Applications Chapter 23: Museum, Galleries, Archives, and Libraries. American Society of Heating, Refrigerating and Air-Conditioning Engineers, SI Edition; ASHRAE: Atlanta, GA, USA. **2015**

[115] DE UNIFICAZIONI, U.-E.N.I. UNI 10829-Works of Art of Historical Importance-ambient Conditions for the Conservation-measurements and Analysis. Milão: UNI; 1999

[116] Ascione F, Bellia L, Capozzoli A. A coupled numerical approach on museum air conditioning: Energy and fluid-dynamic analysis. Applied Energy. 2013;**103**:416–427

[117] Luo X, Chang B, Tian W, Li J, Gu Z. Experimental study on local environmental control for historical site in archaeological museum by evaporative cooling system. Renewable Energy. 2019;**143**:798–809

[118] Chang B, Dang Y, Luo X, Yu CW, Gu Z. Sustainability of evaporative cooling system for environment control for preservation of unearthed historical sites within archaeological museums in China. Sustainability. 2020;**12**:9882

[119] Brahmanis A, Lešinskis A. Indirect evaporative pre-cooled compressor cooling system performance under various outdoor air humidity conditions. In Proceedings of Conference "Civil Engineering; p. 5

[120] Wan Y, Huang Z, Soh A, Chua KJ. On the performance study of a hybrid indirect evaporative cooling and latent-heat thermal energy storage system under commercial operating conditions. Applied Thermal Engineering. 2023;**221**:119902

[121] Isetti C, Lazzari S. Enhancing the energy efficiency of land vehicles air conditioning units through a new evaporative membrane cooler. Applied Thermal Engineering. 2024;**247**:123031

[122] Qi Z, Zhao Y, Chen J. Performance enhancement study of mobile air conditioning system using microchannel heat exchangers.

International Journal of Refrigeration. 2010;**33**:301–312

[123] Filina-Dawidowicz L, Filin S, Wojnicz L, Miłek D, Grzelak P. Energy-efficient maritime transport of refrigerated containers. Procedia Computer Science. 2022;**207**:3572–3581

[124] Chen Y, Luo Y, Yang H. Fresh air pre-cooling and energy recovery by using indirect evaporative cooling in hot and humid region–a case study in Hong Kong. Energy Procedia. 2014;**61**:126–130

Chapter 2

Energy Analysis of Control Methods in a Dual-Pump Direct-Driven Hydraulic System

Juraj Benić and Danko Brezak

Abstract

Direct-driven hydraulic (DDH) systems reduce the inefficiencies of conventional hydraulics by converting electrical energy into hydraulic power on demand via electric servo-driven pumps. To evaluate and minimize irreversibilities in this dual-pump system, several control methods were analyzed with respect to energy efficiency. Proportional–integral–derivative control, linear–quadratic regulation with integral action, sliding-mode control, and backstepping control were implemented and tested under controlled laboratory conditions using a fully loaded cylinder to replicate real operating scenarios. Energy efficiency was quantified by computing each component's power consumption from measured data, yielding overall system efficiency and revealing component-level irreversibilities that pinpoint the key factors affecting performance. A comparative evaluation of the control strategies highlights their relative abilities to lower input power and irreversibilities while preserving stability and performance. Additionally, the trade-off between system dynamics and energy efficiency was explored, showing how control precision affects power usage. Understanding these interactions is essential for optimizing DDH systems in industrial applications.

Keywords: direct-driven hydraulics, energy efficiency, position control, PID control, linear quadratic regulator, sliding mode control, backstepping method

1. Introduction

Electrohydraulic servo systems are used in many industrial applications because they combine the ability to achieve high power through hydraulic means with the ease of processing and transmitting control signals using electrical components [1]. They offer high specific power (a favorable power-to-weight ratio), efficient heat dissipation through hydraulic oil, and simultaneous component lubrication. Hydraulic actuators can generate high forces and torques while delivering fast response times and directional changes, simple overload protection, and high system accuracy [2, 3]. These systems can be easily controlled using valves and pumps, allowing for continuous operation without the risk of overload or damage. As a result, they are considered one of the best solutions for applications requiring rapid and precise control, such as aircraft, satellites, flight simulators, turbine control, heavy mobile machinery, and military equipment.

However, despite these advantages, hydraulic systems also have drawbacks. The need for extremely tight tolerances makes component manufacturing expensive. Additionally, hydraulic systems exhibit highly nonlinear and complex dynamics due to fluid compressibility, the relationship between flow and pressure in the valve (including pressure drops caused by fluid throttling), friction, internal and external fluid leakage, and parameter variations during system operation [4]. These factors pose significant challenges in control system design.

Furthermore, hydraulic energy is not as readily available as electrical energy. Compared to mechanical and electromechanical systems, hydraulic systems generate the most heat and have the lowest energy efficiency. Energy efficiency has become crucial in all fields of mechanical engineering, partly due to strict regulations imposed by authorities to significantly reduce greenhouse gas emissions, particularly CO_2 emissions [5]. Lower efficiency results in higher energy consumption and increased exhaust gas emissions when hydraulic systems are integrated with internal combustion engines. This, in turn, leads to higher operating costs, the need for larger pumps, and more complex cooling systems for heat dissipation.

As a result, various energy-efficient hydraulic systems have been developed over the past two decades to replace the currently most widely used electrohydraulic servo systems, which are controlled solely by servo and proportional valves. These include electrohydraulic actuators, load-sensing systems, direct-driven hydraulic systems (DDH), and others [6–8]. These systems incorporate advanced hydraulic technology, replacing traditional valve-controlled systems with digitally controlled pump-driven solutions. Compared to traditional systems, pump-driven hydraulic systems offer a more compact structure, enhanced reliability, greater precision, and improved energy efficiency by eliminating throttling losses.

Although some of the aforementioned systems may be partially valve-driven, DDH systems are fully pump-driven, making them one of the most advanced and energy-efficient types of hydraulic systems. Numerous studies have been conducted on the design of DDH systems, including applications in marine systems [9], hybrid excavators [10], backhoes [11], pipe-bending systems [12], and various test beds designed to emulate real industrial applications [13].

In all these studies, as well as many others, the integration of DDH systems into existing hydraulic systems was examined. The energy efficiency of these systems was analyzed, pressure drops within the system were studied, and component heating was investigated. Additionally, motor and generator operating modes were presented, along with the potential for energy regeneration in generator mode.

Unlike previous studies that utilized a single-pump DDH system, typically with variable-displacement pumps (pumps with variable specific volume), this research employs a newer type of system proposed in 2014 [14], in which a cylinder is controlled by two fixed-displacement pumps (pumps with a constant specific volume) connected to a common shaft. A simplified schematic representation of the system is shown in **Figure 1**.

These two bidirectional fixed-displacement pumps are directly driven by a servo motor, with pressure, flow, and fluid direction regulated by adjusting the pumps' speed and rotation direction. As a result, various servo and/or proportional valves for actuator speed and direction control are no longer required.

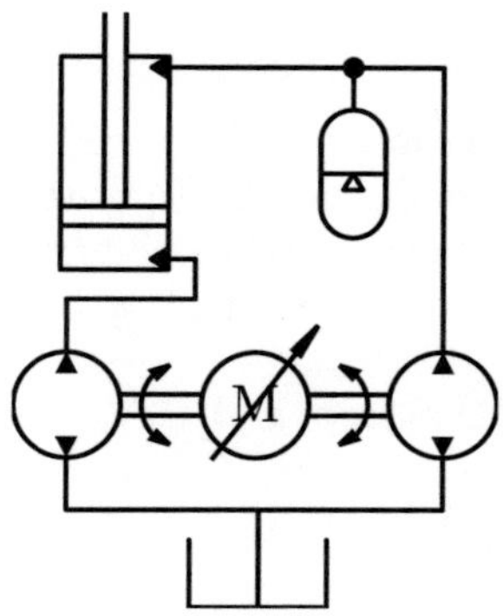

Figure 1.
Simplified schematic representation of the DDH system according to Minav et al. [14].

Additionally, fixed-displacement pumps are significantly cheaper than variable-displacement pumps, making this dual-pump DDH system a more cost-effective and energy-efficient alternative.

The initially proposed system was further modified by replacing the hydraulic reservoir with a single hydraulic accumulator, resulting in a more compact structure that occupies minimal space [15]. In [16], the system's design and component selection were described in detail. A leakage model for the pumps in the compact system version was developed, along with a comprehensive model of irreversible energy losses (irreversibilities) derived from experimental data. The findings revealed that the majority of losses – approximately 33.5% – were due to hydromechanical losses, while the system's output potential energy accounted for around 45.8%.

A thermal analysis of this DDH system was conducted in Minav et al. (2016) [17], indicating that the hydraulic components operated within safe temperature limits, with the highest temperature recorded in the electric motor. Additionally, the energy efficiency of the proposed DDH system was analyzed in several studies [18–20], demonstrating that it ranks among the most energy-efficient solutions within pump-controlled hydraulic systems.

The implementation of DDH systems has been widely studied in various applications. Research has focused on assessing their energy efficiency [14], analyzing pressure losses within the system [21], and examining the heating effects of components [17]. Furthermore, studies have explored motor and generator operating modes [22], including the potential for energy regeneration in generator mode [23].

In the case of the previously mentioned DDH system with two bidirectional pumps of constant specific volume, research has primarily focused on system (hardware) design and energy efficiency. However, energy efficiency has mostly been analyzed from the perspectives of the selection and configuration of physical components, temperature effects, and oil type. This suggests that the impact of control strategies on energy efficiency in such advanced electrohydraulic systems remains understudied, emphasizing the need for further study and analysis. One such study is presented in this chapter.

A detailed description of the analyzed system and control algorithms, the experiment, the obtained results, and the discussion is presented in the subsequent sections.

2. Experimental setup

For the purpose of testing various control algorithms, comparing energy efficiency, and analyzing system dynamics of pump-driven and valve-driven electrohydraulic systems, an experimental setup was designed and built, as shown in **Figure 2**. The setup includes two independent systems:

- A pump-controlled (DDH) system and
- A proportionally valve-driven system.

Both systems share a common 15-liter oil reservoir (containing Castrol Hyspin AWS 32 mineral oil) and are connected in parallel to a vertically positioned actuator, allowing for a direct performance comparison under identical loading conditions. The schematic representation of the test bed structure is shown in **Figure 3**.

A double-acting custom-made hydraulic cylinder ($\phi 32/22 \times 300$ mm) manufactured by the Hidromehanika Ltd. was used as the actuator. Electromagnetic 2/2 valves (Argo-Hytos Co., model SD2E-A2) are used to bypass the flow sensors in tests where oil flow measurement is not required. An electromagnetic 4/2 valve

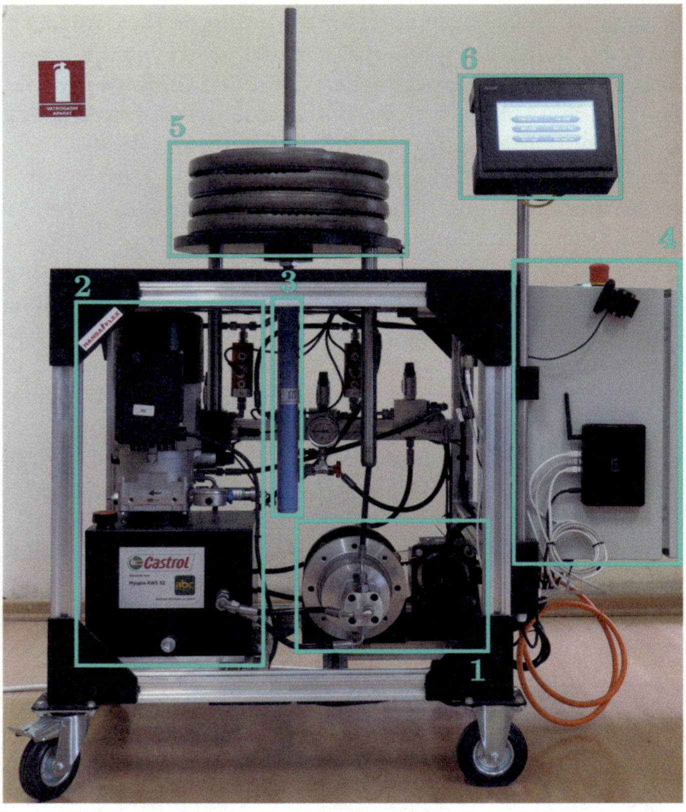

Figure 2.
Hydraulic test setup: (1) DDH system, (2) proportional electrohydraulic system, (3) cylinder, (4) control cabinet, (5) weights, and (6) HMI [24].

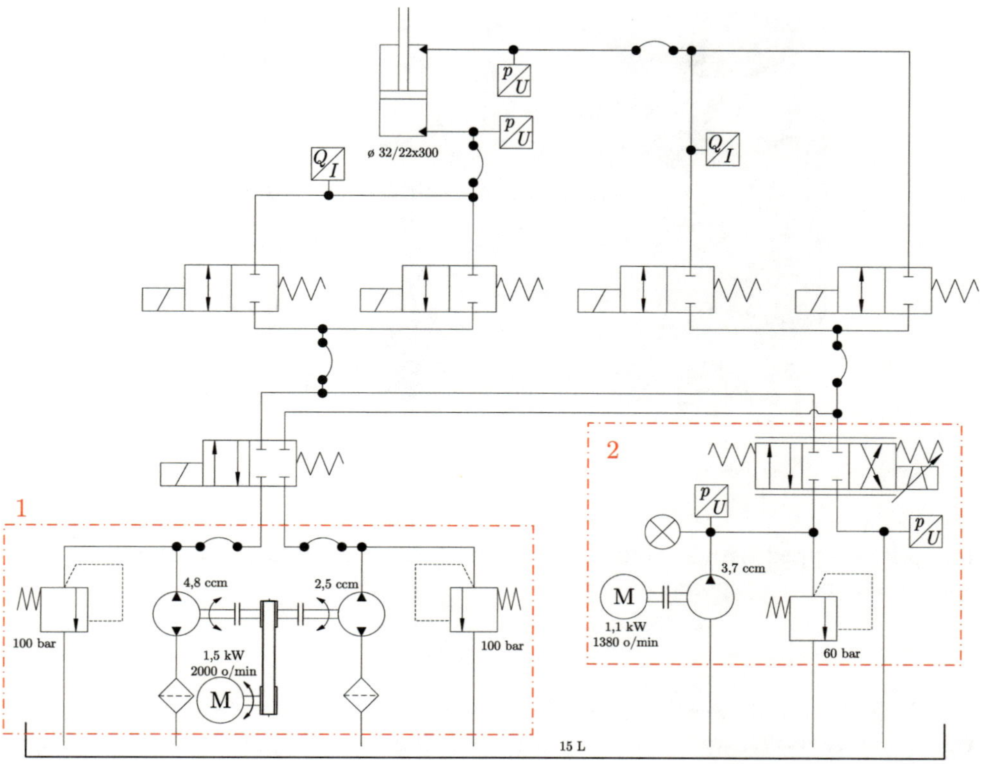

Figure 3.
Hydraulic circuit: (1) DDH system and (2) proportional electrohydraulic system [24].

(Argo-Hytos Co., model SD2E-A4) is used to activate the DDH system. Pressure sensors are connected to the chambers of the double-acting cylinder, the oil reservoir, and the pump outlet in the proportional system, while flow sensors are installed at the cylinder chamber inlets. A detailed description of the measuring devices is provided in Section. 2.2.

Since this research was based on the analysis of the energy efficiency of the DDH system, the proportional electrohydraulic system was not used.

2.1 Direct-driven hydraulic system

The DDH system is shown in **Figure 4**.

The system is controlled by an AC servo motor (Mitsubishi, model HG-SN125JK), which is speed-controlled by a servo drive. Power is transmitted to the pumps via a belt transmission, enabling direct control of the cylinder movement. The system employs two reversible gear pumps (Argo-Hytos, model GP1), which are coupled via a shaft to directly regulate oil flow to the cylinder.

On the piston side (A-side), a GP1-4.8 pump with a specific volume of 4.8 cm^3 supplies flow into the cylinder, while simultaneously, a GP1-2.5 pump with a specific volume of 2.5 cm^3 manages the output flow from the cylinder (B-side). The input and output flows depend on both the direction and speed of pump rotation. The pressure required for load movement is determined by the load's magnitude. Each

Figure 4.
DDH system: (1) A-side pump, (2) B-side pump, and (3) servo motor [24].

pump is equipped with a safety valve. Pump selection was based on the cylinder specifications, as proposed in [16]:

$$\frac{A_B}{A_A} = \frac{D_B}{D_A}, \tag{1}$$

where A_A and A_B are the areas of the piston and rod sides of the cylinder, respectively, and D_A and D_B are the specific volumes of the pumps corresponding to these sides.

2.2 Data acquisition and system control

The acquisition of process variables in the experimental setup is enabled through pressure, flow, and position sensors. All sensors used are powered from the control cabinet with a supply voltage of 24 VDC. Pressure measurements in the cylinder chambers and oil reservoir are performed using Hydac pressure sensors, model HDA 7446. The measuring range of the sensors is from 0 to 100 bar, with a linear voltage output characteristic ranging from 0 to 10 VDC. To measure the volumetric oil flow into the cylinder chambers, Hydac flow sensors, model EVS 3,106, are used. These sensors are calibrated for a specific flow direction and a measurement range of 1.2 to 20 l/min, within which they exhibit a linear characteristic. For flow rates below 1.2 l/min or in the reverse direction, the sensor yields less accurate measurements. The output signal is current-based, in the range of 4 to 20 mA. A Micro-Epsilon industrial displacement sensor, model WDS-500-P60, is used to measure the position of the hydraulic cylinder, while the velocity (v_{est}) is estimated in real time from the position data using the procedure described in [24]. The sensor's working range is 0 to 500 mm, and its output is a voltage signal ranging from 0 to 10 VDC.

The entire system is controlled using a Mitsubishi PLC, model FX5U-32MT/ESS. The PLC features 16 digital inputs, 16 digital outputs, two analog inputs in the range of $0 - 10$ VDC, and one analog voltage output in the same range. Two 14-bit A/D modules and one 14-bit D/A module, each with configurable input/output modes (current or voltage), are connected to the PLC. The A/D modules acquire analog signals from the sensors described above, while the D/A module generates the analog control signal for the servo drive. The PLC's digital outputs are used to control DC relays and indicator lights. The DC relays supply power to the 2/2 valves used for bypassing the flow sensors, and the 4/2 valve used to activate the direct-driven electrohydraulic system. The servo drive regulates the rotational speed of the servo motor and communicates back to the PLC via serial communication, providing real-time data on motor speed, torque, and power output. A router is installed on the outer side of the control cabinet, enabling communication between the PLC, HMI, and server. The HMI is used for local system control and parameter adjustment, while also displaying all key process variables for monitoring and control.

Experimental data were collected for cylinder loads ranging from 0 to 200 kg, in 20 kg increments. Each test lasted 60 seconds, with measurement data logged every 10 milliseconds. Two test signals were used:

- A step signal consisting of six randomly chosen steps (280, 10, 150, 260, 50, and 0 mm), each changing every 10 seconds and
- A sine-wave signal with an amplitude of 0.05 meters, a frequency of 0.2 Hz, and a bias of 0.15 meters.

3. Mathematical modeling

To successfully design various controllers used in this study, it is necessary to derive a mathematical model of the system. For the linear quadratic regulator with integral action (LQR-I), a linearized model is employed to calculate the optimal gains, while a state-space model is obtained to determine the parameters for the backstepping controller. Additionally, the functionality of the proposed controllers is verified through simulation-based testing.

3.1 Model of the hydraulic cylinder

The schematic representation of the asymmetric double-acting hydraulic cylinder and the forces acting on it are shown in **Figure 5**. The objective is to model the actuator responsible for lifting the load in a gravitational field.

From Newton's second law, the force balance equation for the hydraulic cylinder can be written as follows:

$$m\ddot{y} = p_A A_A - p_B A_B - F_r - mg, \tag{2}$$

where p_A and p_B are the pressures inside the cylinder chambers, F_r is the friction force, g is the acceleration due to gravity, and m is the mass of the load.

The LuGre friction model, widely accepted in hydraulics, will be used to represent the friction force. This model extends Dahl's model and also describes the Stribeck effect and the stick-slip phenomenon [25]. A detailed description of the proposed friction model and its characteristics is provided in Olsson (1996) [26], and it is defined by the following equations:

$$F_r = \sigma_0 z + \sigma_1 \dot{z} + \sigma_2 v, \tag{3}$$

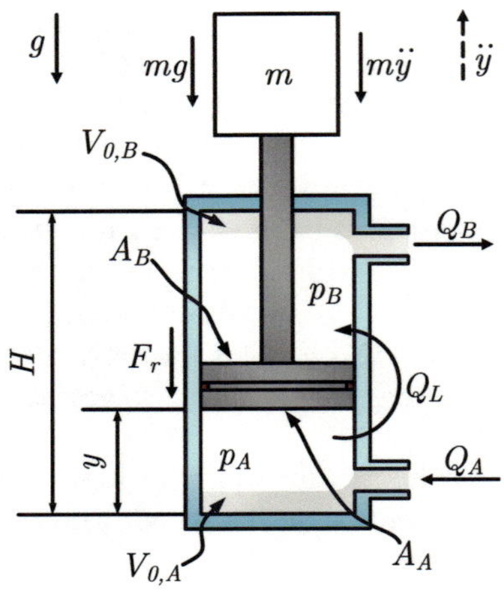

Figure 5.
Schematic representation of the hydraulic cylinder.

$$\dot{z} = v - \frac{\sigma_0 z}{g(v)}|v|, \tag{4}$$

$$g(v) = F_c + (F_s - F_c)e^{-\left(\frac{v}{v_s}\right)^2}, \tag{5}$$

where v is the relative velocity between the surfaces (equal to the cylinder velocity $\dot{y}$), F_c and F_s represent Coulomb friction and static friction, v_s is the Stribeck velocity, σ_2 is the viscous friction coefficient, and z is the state variable representing the average deflection of the bristles. Additionally, σ_0 is the bristle-spring constant, and σ_1 is the bristle-damping coefficient.

Pressure dynamics of the cylinder chambers are given by Eq. (1):

$$\dot{p}_A = \frac{B}{V_{0,A} + A_A y}(Q_A - Q_L - Q_E - A_A \dot{y}), \tag{6}$$

$$\dot{p}_B = \frac{B}{V_{0,B} + A_B(H - y)}(Q_B + Q_L - Q_E + A_B \dot{y}), \tag{7}$$

where $\dot{p}_A$ and $\dot{p}_B$ represent the pressure variations in the cylinder chambers, $V_{0,A}$ and $V_{0,B}$ are the dead volumes, y is the position and $\dot{y}$ is the cylinder velocity, H is the maximum stroke of the cylinder, Q_A and Q_B are the flow rates, and Q_L and Q_E denote internal and external leakage, respectively. Internal leakage and external leakage have a negligible effect on the cylinder dynamics and, according to Ruderman [27], can be neglected.

3.2 Model of the DDH system

To describe the system dynamics, the state-space form of a model is used. The proposed formulation of an nth-order dynamic system reduces it to n mutually coupled first-order differential equations, and in the general case, is given as follows:

$$\dot{\mathbf{x}}(t) = \mathbf{f}\left(\mathbf{x}(t), \mathbf{u}(t), t\right), \tag{8}$$

where $\mathbf{x}(t) \in \mathrm{R}^n$ is the state variable vector, $\mathbf{u}(t) \in \mathrm{R}^n$ is the system input vector, and $\mathbf{f}(å, å, t)$ is a nonlinear vector function that describes the system dynamics.

From Eqs. (6), (7), (2), and (4), the DDH system model in state-space form, according to Expression (8), can be written as follows:

$$\dot{x}_1 = x_2,$$

$$\dot{x}_2 = \frac{x_4 A_A - x_5 A_B - F_r - mg}{m},$$

$$\dot{x}_3 = x_2 - \frac{\sigma_0 x_3}{g(x_2)}|x_2|, \tag{9}$$

$$\dot{x}_4 = \frac{B}{V_{0,A} + A_A x_1}(Q_{p,A} - A_A x_2),$$

$$\dot{x}_5 = \frac{B}{V_{0,B} + A_B(H - x_1)}(Q_{p,B} + A_B x_2),$$

where the state variables $\mathbf{x} = [x_1, x_2, x_3, x_4, x_5]^T$ are defined as follows: $x_1 = y$, $x_2 = \dot{y}$, $x_3 = z$, $x_4 = p_A$, and $x_5 = p_B$. The system inputs are the flow rates $Q_{p,A}$ and $Q_{p,B}$ given as follows:

$$Q_{p,A} = \omega D_A \eta_v,$$

$$Q_{p,B} = -\omega D_B \eta_v, \tag{10}$$

where ω is the angular velocity of the servo motor, and η_v is the efficiency of the pumps. Furthermore, the angular velocity of the motor can be related to the input voltage signal as follows:

$$\omega = K_{\omega,u} u, \tag{11}$$

where $K_{\omega,u}$ is the speed constant that converts the input voltage at the servo drive into the rotational speed of the servo motor, and u is the control variable. To complete the simulation model, two additional expressions are required: Eq. (3) for calculating the friction force and Eq. (5) for computing the function $g(v)$.

4. Control methods for the DDH system

Classical approaches to hydraulic system control rely on linearizing system dynamics around an operating point and applying controllers with constant gain values within a closed-loop framework [28, 29]. However, this method does not ensure satisfactory performance in precise motion control tasks when system operating conditions change. Due to the inherent limitations of classical controllers, adaptive control algorithms have been introduced. Self-tuning control algorithms enhance reference trajectory tracking by utilizing information from previous motions [30, 31], while model reference adaptive control algorithms improve system robustness against parameter variations during operation [32, 33].

A major drawback of adaptive controllers is their reliance on accurate estimation of the system's structure and parameters, which requires a deep understanding of system dynamics. Given the highly nonlinear nature of hydraulic systems – characterized by parameter variability due to component properties, disturbances, and load fluctuations during operation [34–36] – deriving a mathematical model that precisely captures system behavior is challenging. This issue becomes even more pronounced in complex mechanical structures, such as robotic systems with hydraulic actuation [31].

Existing adaptive controllers still rely on linearized system models, making them robust only to minor parameter variations and ensuring only local stability. In contrast, variable structure control and sliding mode control (SMC) methods offer robustness to more significant parameter variations [37–39]. However, their practical implementation presents challenges, particularly in selecting appropriate sliding surface parameters and mitigating the adverse effects of high-frequency control signal oscillations (chattering).

To address nonlinearities and unmodeled system dynamics, nonlinear control strategies based on Lyapunov stability analysis have been proposed. One such approach is backstepping control, which enables improved performance in both transient and steady-state conditions [40, 41].

Taking this into account, the energy efficiency of the analyzed dual-pump DDH system was evaluated under four control strategies:

- Two linear methods – the standard industry-adopted–(proportional–integral–derivative) (ISA–PID) controller and the LQR-I, and
- Two nonlinear methods – the SMC and the backstepping controller.

The controller parameters were adjusted using a trial-and-error approach, informed by extensive experimental evaluation, to ensure satisfactory performance under varying load conditions.

4.1 ISA–PID controller

In automatic control, the PID controller is one of the most commonly used control algorithms due to its key features, such as providing feedback on the current system state, eliminating steady-state error through integral action, and predicting future system behavior using derivative action. Due to its simple structure, intuitiveness, and ability to achieve satisfactory performance in various processes, the PID controller has become the standard controller in industrial applications. Today, the PID controller is typically implemented in a discrete form and can be found in all types of control equipment, either as a standalone controller or as a function block within a PLC. By combining proportional (P), integral (I), and derivative (D) actions, different PID controller structures can be obtained. The most general structure is the ISA–PID controller, which is given by the following expression [42]:

$$u = K_p(\beta y_r - y_m) + K_i \int_0^t e(\tau)d\tau + K_d \frac{d}{dt}(\gamma y_r - y_m), \tag{12}$$

where K_p, K_i, and K_d are the proportional, integral, and derivative gains of the controller, respectively, y_r and y_m are the desired and measured cylinder positions, and β and γ are weighting factors for the reference cylinder position, ranging within $0 \leq \beta, \gamma \leq 1$.

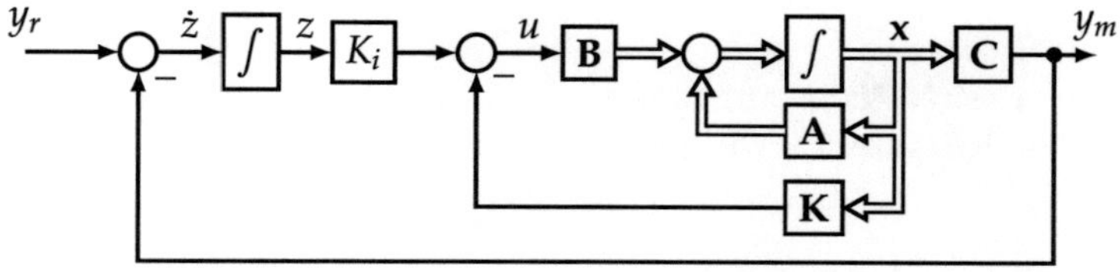

Figure 6.
LQR controller with integral action.

4.2 LQR-I controller

The LQR controller with integral action (LQR-I) is an extension of the LQR controller and was used in Pourebrahim et al. (2016) [43] for tracking the desired trajectory of a wet clutch actuated by a hydraulic servo system. The proposed controller proved to be simple and fast for implementation in an electrohydraulically actuated wet clutch. The structure of the proposed LQR-I controller is shown in **Figure 6**.

Integral action is incorporated into the LQR controller to enable trajectory tracking in such a way that the steady-state tracking error is eliminated. The fundamental approach of this type of controller is to create a new state within the controller, which computes the integral of the tracking error and is then used to extend the existing linearized state-space system model obtained from Eq. (9). The integral state z is defined as follows:

$$\dot{z} = y_r - y_m = y_r - \mathbf{C}\mathbf{x}, \tag{13}$$

where $\dot{z}$ is the integral state, and $\mathbf{Cx}$ represents the system output to be regulated. By incorporating the integral state from Eq. (13) into the linearized state-space system model, we obtain a new state-space system in the following form:

$$\begin{aligned} \dot{\widetilde{\mathbf{x}}} &= \widetilde{\mathbf{A}}\widetilde{\mathbf{x}} + \widetilde{\mathbf{B}}u + \widetilde{\mathbf{E}}y_r, \\ y &= \widetilde{\mathbf{C}}\widetilde{\mathbf{x}}, \end{aligned} \tag{14}$$

where

$$\dot{\widetilde{\mathbf{x}}} = \begin{bmatrix} \mathbf{x} \\ \dot{z} \end{bmatrix}, \quad \widetilde{\mathbf{A}} = \begin{bmatrix} \mathbf{A} & \mathbf{0} \\ -\mathbf{C} & 0 \end{bmatrix}, \quad \widetilde{\mathbf{x}} = \begin{bmatrix} \mathbf{x} \\ z \end{bmatrix}, \quad \widetilde{\mathbf{B}} = \begin{bmatrix} \mathbf{B} \\ 0 \end{bmatrix}, \quad \widetilde{\mathbf{E}} = \begin{bmatrix} 0 \\ I \end{bmatrix}, \quad \widetilde{\mathbf{C}} = [\mathbf{C} \quad 0].$$

The newly extended state-space matrices are used to compute the optimal controller gain, where the control law is given as follows:

$$u = -\widetilde{\mathbf{K}}\,\widetilde{\mathbf{x}} = -\mathbf{K}\,\mathbf{x} - K_i z, \tag{15}$$

where $\widetilde{\mathbf{K}}$ is the vector of optimal gains $[\mathbf{K}\,K_i]$.

4.3 Sliding mode controller

The SMC is a robust and widely studied nonlinear control strategy, particularly in recent years. This approach is known for its high accuracy, strong robustness, and straightforward implementation with easy parameter tuning. In this work, a PD-based SMC controller is employed, with its sliding surface defined as follows [44]:

$$s = \dot{e} + \lambda e, \tag{16}$$

where λ is a positive constant that can be interpreted as the slope of the sliding surface, and e is the system error given by the following:

$$e = y_r - y_m. \tag{17}$$

The control law is defined as follows:

$$u = k_1 s + k_2 \text{sign}(s), \tag{18}$$

where k_1 and k_2 are the controller gains, and the sign function is defined as follows:

$$\text{sign}(x) = \frac{x}{|x| + \grave{o}}, \tag{19}$$

where $\grave{o}$ is a very small number used to smooth the function and reduce chattering.

5. Backstepping controller

Designing a control law using the Lyapunov function is a valuable technique in control theory. However, determining an appropriate Lyapunov function for general nonlinear system control can be challenging. A control law design based on Lyapunov stability analysis can be systematically addressed using a recursive approach known as the backstepping method. This proposed method is particularly suitable for strict-feedback systems, which can be expressed in the following form [45]:

$$\begin{aligned}
\dot{\xi}_1 &= f_1(\xi_1) + g_1(\xi_1)\xi_2, \\
\dot{\xi}_2 &= f_2(\xi_1, \xi_2) + g_2(\xi_1, \xi_2)\xi_3, \\
&\vdots \\
\dot{\xi}_{r-1} &= f_{r-1}(\xi_1, \xi_2, \dots, \xi_{r-1}) + g_{r-1}(\xi_1, \xi_2, \dots, \xi_{r-1})\xi_r, \\
\dot{\xi}_r &= f_r(\xi_1, \xi_2, \dots, \xi_r) + g_r(\xi_1, \xi_2, \dots, \xi_r)u,
\end{aligned} \tag{20}$$

where $\xi_1, \dots, \xi_r \in R$ and $u \in R$ are the control inputs, while f_i and g_i for $i = 1, \dots, r$ are known functions.

In order to apply the backstepping method to the system in Eq. (9), the system's dynamic model must be expressed in the form of a strict-feedback model, as given by Eq. (20). According to Assegu et al. (2014) [46], this can be achieved by defining a new state variable that represents the output force of the hydraulic cylinder as follows:

$$\overline{x}_4 = x_4 A_A - x_5 A_B, \tag{21}$$

From Eq. (9), by applying Eq. (21), the state-space model is given as follows:

$$\begin{aligned}
\dot{x}_1 &= x_2, \\
\dot{x}_2 &= \frac{\overline{x}_4 - F_r - mg}{m}, \\
\dot{x}_3 &= x_2 - \frac{\sigma_0 x_3}{g(x_2)}|x_2|, \\
\dot{\overline{x}}_4 &= \gamma_1 u - \gamma_2 x_2,
\end{aligned} \tag{22}$$

where γ_1 and γ_2 are given as follows:

$$\gamma_1 = \frac{A_A B}{V_{0,A} + A_A x_1} Q_{p,A} + \frac{A_B B}{V_{0,B} + A_B (H - x_1)} Q_{p,B},$$
$$\gamma_2 = \frac{A_A^2 B}{V_{0,A} + A_A x_1} + \frac{A_B^2 B}{V_{0,B} + A_B (H - x_1)}. \tag{23}$$

The backstepping method will be applied to the system described by Eq. (22). For the given system, the negative tracking errors corresponding to each state variable directly influenced by the control input are defined as follows:

$$e_i = x_i - x_{id}, \tag{24}$$

where x_{id} is the desired value of the state variable, and $i = 1, 2, 4$. The third state variable is omitted, as it is not directly related to the input–output path.

The first step is to determine the virtual control variable x_{2d} based on the Lyapunov function of the first error variable, which is defined as follows:

$$V_1 = \frac{1}{2} e_1^2, \tag{25}$$

and its derivation is given as follows:

$$\dot{V}_1 = e_1 \dot{e}_1 = e_1 (x_2 - \dot{x}_{1d}). \tag{26}$$

To stabilize the given subsystem, it is necessary to select an appropriate virtual control variable x_{2d}. The first intermediate control law can be defined from Eq. (26) as follows:

$$x_{2d} = -k_1 e_1 + \dot{x}_{1d}. \tag{27}$$

By substituting the given control law and introducing the second error variable (e_2) into Eq. (26), we obtain:

$$\dot{V}_1 = e_1 (e_2 + x_{2d} - \dot{x}_{1d}) = -k_1 e_1^2 + e_1 e_2, \tag{28}$$

where k_1 and e_2 will be defined later to ensure that the condition $\dot{V}_1 < 0$ is satisfied.

The second step is to determine the virtual control variable $\overline{x}_{4d}$ based on the Lyapunov function of the second error variable, defined as follows:

$$V_2 = V_1 + \frac{1}{2} e_2^2, \tag{29}$$

and its derivation is given as follows:

$$\dot{V}_2 = \dot{V}_1 + e_2 \dot{e}_2 = -k_1 e_1^2 + e_1 e_2 + e_2 (\dot{x}_2 - \dot{x}_{2d})$$
$$= -k_1 e_1^2 + e_1 e_2 + e_2 \left(\frac{\overline{x}_4}{m} - \frac{F_r}{m} - g - \dot{x}_{2d} \right). \tag{30}$$

If we define the fourth error variable as:

$$e_4 = \overline{x}_4 - \overline{x}_{4d}, \tag{31}$$

then the intermediate control law $\overline{x}_{4d}$ is given as follows:

$$\overline{x}_{4d} = m\left(\frac{F_r}{m} + g + \dot{x}_{2d} - k_2 e_2 - e_1\right). \tag{32}$$

By substituting Eqs. (31) and (32) into Eq. (30), the derivative of the Lyapunov function simplifies to the following:

$$\dot{V}_2 = -k_1 e_1^2 - k_2 e_2^2 + \frac{1}{m} e_2 e_4. \tag{33}$$

The control law for global asymptotic stability of the system will be derived from the Lyapunov function:

$$V_3 = V_2 + \frac{1}{2} e_4^2, \tag{34}$$

whose derivative is given by the following:

$$\dot{V}_3 = -k_1 e_1^2 - k_2 e_2^2 + \frac{1}{m} e_2 e_4 + e_4(\gamma_1 u - \gamma_2 x_2 - \dot{\overline{x}}_{4d}). \tag{35}$$

From Eq. (35), the control law is chosen as follows:

$$u = \frac{1}{\gamma_1}\left(\gamma_2 x_2 + \dot{\overline{x}}_{4d} - k_3 e_4 - \frac{1}{m} e_2\right). \tag{36}$$

For global system stabilization, the Lyapunov function given by Eq. (35), together with the proposed control law, must satisfy the following condition:

$$\dot{V}_3 = -k_1 e_1^2 - k_2 e_2^2 - k_3 e_4^2 \le 0. \tag{37}$$

It follows that the gains must satisfy $k_1, k_2, k_3 > 0$ to ensure that the derivatives of the Lyapunov functions are negative semi-definite. This guarantees global asymptotic tracking using the control laws given by Eqs. (27), (32), and (36).

5.1 System error metrics

To evaluate the performance of the control strategies, the integral of squared error (ISE) is used as the primary metric for quantifying tracking accuracy under both step and sine-wave reference signals. This metric emphasizes the magnitude of the tracking error over time and penalizes larger deviations more heavily [47]:

$$\text{ISE} = \int e(t)^2 dt. \tag{38}$$

The ISE metric is used to compare the effectiveness of the four control methods – PID, LQR-I, SMC, and backstepping – in minimizing tracking errors across different testing scenarios.

6. Energy efficiency

6.1 Power definitions for DDH system

To determine the overall efficiency of the systems, the power of each component is calculated based on measured data and estimated for components where direct measurement is not possible. The input power of the hydraulic cylinder is given by the following [21]:

$$P_{cyl} = (p_A A_A - p_B A_B) v_{est}, \tag{39}$$

where p_A and p_B are the measured pressures on the piston and rod sides, respectively, and v_{est} is the estimated cylinder velocity. The output power of each pump is calculated as follows:

$$P_{pump} = Qp, \tag{40}$$

where Q is the pump's output flow (Q_A and Q_B) and p is the measured supply pressure (p_A and p_B). The input electrical power P_{el} of the DDH system is directly measured and obtained from the servo drive. The servo motor output power is calculated from the measured torque and angular velocity of the servo motor shaft as follows [21]:

$$P_{motor} = T\omega, \tag{41}$$

where T is the servo motor torque and ω is the rotational speed.

6.2 Analysis of energy efficiency

The average energy efficiency of the system is defined as follows:

$$\eta = \frac{E_{out}}{E_{in}} = 1 - \frac{E_{loss}}{E_{in}}, \tag{42}$$

where

$$E_{in} = \int P_{el}, dt + \int |P_{cyl}| \Big|_{Pcyl < 0} dt, \text{ and} \tag{43}$$

$$E_{out} = \int P_{cyl} \Big|_{Pcyl \geq 0} dt \tag{44}$$

stands for the input energy or total power supplied to the system and the output energy or the useful mechanical power delivered by the cylinder, respectively. To accurately estimate energy efficiency, the input and output energy are calculated based on the power flow direction. For lifting movements of the cylinder, the input power corresponds to the electrical power supplied to the system. For lowering movements, an additional input must be considered in the form of the energy contributed by the load performing work on the system (i.e., when the cylinder is pushed by the load). The E_{loss} quantifies the irreversible energy losses

$$E_{\text{loss}} = E_{\text{in}} - E_{\text{out}}, \tag{45}$$

which can be decomposed into the following components:

$$E_{\text{loss}} = E_{\text{loss,motor}} + E_{\text{loss,pumps}} + E_{\text{loss,hydraulic}}. \tag{46}$$

$$E_{loss,motor} = \int P_{el} dt - \int P_{motor} dt.$$

$$E_{loss,pumps} = \int P_{motor}dt - \left(\int P_{pump,1}dt - \int P_{pump,2}dt \right), \quad (47)$$

$$E_{loss,hydraulic} = \left(\int P_{pump,1}dt - \int P_{pump,2}dt \right) - E_{out},$$

where

• $E_{loss,motor}$ represents irreversible energy losses during conversion from the servo drive's electrical input to the servo motor's mechanical output,

• $E_{loss,pumps}$ represents the irreversible energy losses in the pump assembly during mechanical-to-hydraulic energy conversion, and

• $E_{loss,hydraulic}$ represents the irreversible energy losses due to hydraulic irreversibilities (viscous friction, throttling, leakage) between the pumps and the hydraulic cylinder.

Irreversible energy losses have been expressed as percentages and compared in the following section using their normalized form:

$$L_{motor} = \frac{E_{loss,motor}}{E_{in}} \times 100\%,$$

$$L_{pumps} = \frac{E_{loss,pumps}}{E_{in}} \times 100\%, \quad (48)$$

$$L_{hydraulic} = \frac{E_{loss,hydraulic}}{E_{in}} \times 100\%,$$

while the average energy efficiency from Eq. (42) has also been expressed as a percentage as follows:

$$\eta(\%) = \eta \times 100\%. \quad (49)$$

6.3 Results

The evaluation of energy efficiency and tracking accuracy for the tested controllers – PID, LQR-I, SMC, and backstepping – under both step and sine-wave reference trajectories reveals several important trends.

As shown in **Tables 1** and **2**, energy efficiency increases consistently with load across all controllers. This is expected because increasing load raises the proportion of useful mechanical output relative to the system's internal irreversibilities. Additionally, under sine-wave tracking, the system operates in a more continuous, dynamic regime, which reduces abrupt transients and lowers irreversibilities, thereby improving overall system efficiency. This leads to slightly higher overall efficiencies compared to the step-input case.

Among all controllers, PID exhibits the highest efficiency at full load, reaching 28.58% under sine-wave tracking, followed closely by LQR-I (27.22%), backstepping (27.61%), and SMC (27.51%). Although the differences are small, the PID controller

Controller	Mass (kg)	η (%)	$L_{hydraulic}$ (%)	L_{motor} (%)	L_{pumps} (%)	ISE
PID	0	2.76	2.31	26.38	68.55	2.29×10^{-7}
	60	7.05	7.96	22.83	62.16	2.20×10^{-7}
	120	10.92	16.74	19.06	53.28	1.99×10^{-7}
	200	13.82	25.50	23.49	37.19	1.94×10^{-7}
LQR-I	0	1.65	2.58	19.84	75.92	2.41×10^{-7}
	60	6.18	7.66	19.21	66.95	2.22×10^{-7}
	120	9.13	15.16	17.57	58.14	2.00×10^{-7}
	200	10.92	24.61	16.23	48.24	1.96×10^{-7}
SMC	0	1.56	2.99	18.07	77.39	2.81×10^{-7}
	60	5.82	8.75	16.54	68.89	2.43×10^{-7}
	120	8.83	17.99	15.49	57.69	2.21×10^{-7}
	200	11.03	26.77	17.01	45.19	1.96×10^{-7}
Backstepping	0	1.50	2.90	19.51	76.09	2.86×10^{-7}
	60	5.62	8.78	16.89	68.70	2.55×10^{-7}
	120	8.84	17.47	16.61	57.08	2.17×10^{-7}
	200	11.09	26.77	16.79	45.35	1.97×10^{-7}

Table 1.
Energy efficiency and irreversible energy losses during step-reference tracking.

consistently offers the best trade-off between irreversibilities and tracking accuracy. Its improved efficiency at higher loads arises from a more favorable ratio of useful mechanical work to internal irreversibilities, particularly within the pump subsystem.

Across all control methods, irreversibilities are similarly distributed:

- The greatest share of irreversibilities occurs in the pumps, especially under low-load conditions.

- Irreversible energy losses in the servo drive motor subsystem remain moderate and relatively stable.

- Hydraulic irreversibilities rise with increasing load but constitute the smallest fraction of total system irreversibilities.

Notably, LQR-I achieves comparable efficiency to PID, despite its more complex implementation and tuning. On the other hand, SMC and backstepping, while nonlinear and generally more robust, tend to exhibit slightly lower efficiency at low loads, though they catch up at higher loads, where system dynamics are more demanding.

In terms of control accuracy, all four controllers demonstrate similar performance, with only minor differences in ISE values across both reference types. This indicates that all methods are capable of delivering stable and precise

Controller	Mass (kg)	η (%)	$L_{hydraulic}$ (%)	L_{motor} (%)	L_{pumps} (%)	ISE
PID	0	4.75	0.98	22.08	72.19	3.27×10^{-8}
	60	16.59	5.34	18.25	59.82	3.18×10^{-8}
	120	24.66	9.37	16.56	49.41	3.29×10^{-8}
	200	28.58	14.55	21.00	35.87	3.09×10^{-8}
LQR-I	0	4.09	1.24	21.32	73.36	3.62×10^{-8}
	60	17.05	6.66	20.30	55.99	3.27×10^{-8}
	120	23.30	9.71	17.31	49.68	3.32×10^{-8}
	200	27.22	12.68	20.92	39.18	3.63×10^{-8}
SMC	0	3.86	1.37	22.27	72.50	3.46×10^{-8}
	60	16.51	6.26	20.46	56.77	3.37×10^{-8}
	120	23.79	10.52	17.58	48.11	3.38×10^{-8}
	200	27.51	13.88	22.69	35.92	3.45×10^{-8}
Backstepping	0	3.88	1.46	22.15	72.50	3.44×10^{-8}
	60	16.63	6.25	17.96	59.15	3.23×10^{-8}
	120	23.36	10.20	18.47	47.97	3.58×10^{-8}
	200	27.61	13.83	22.99	35.58	3.37×10^{-8}

Table 2.
Energy efficiency and irreversible energy losses during sine-wave reference trajectory tracking.

trajectory tracking. At full load, the differences in tracking error become negligible, suggesting robust control behavior across varying operating conditions.

In summary, while all controllers achieve comparable tracking accuracy, PID and LQR-I stand out for their higher energy efficiency, particularly under heavy-load and steady-state conditions. These results suggest that linear controllers better minimize irreversibilities in systems with well-defined dynamics, whereas nonlinear methods, such as sliding-mode control and backstepping, offer higher robustness, especially under uncertainty and nonlinear disturbances.

7. Conclusions

This study comprehensively analyzed the energy efficiency and control performance of a dual-pump DDH system under four distinct control strategies: ISA-PID, LQR-I, SMC, and backstepping. Through rigorous experimentation with both step and sinusoidal reference inputs across a range of load conditions, key insights into the trade-offs between control precision and energy efficiency were obtained.

The results reveal that all controllers achieved satisfactory trajectory tracking performance, with minimal differences in tracking errors as measured by the ISE. However, notable differences emerged in energy efficiency, particularly under heavier loads. Among the evaluated methods, the PID and LQR-I controllers consistently demonstrated higher energy efficiency across operating conditions,

DOI: http://dx.doi.org/10.5772/intechopen.1012744

with PID slightly outperforming the others at full load. In contrast, nonlinear controllers such as SMC and backstepping, while offering strong robustness, exhibited marginally lower efficiency at light loads but matched linear controllers as the load increased.

Additionally, the breakdown of irreversible energy losses shows that pump irreversibilities are dominant, especially at low loads, whereas hydraulic irreversibilities become more significant as the load increases. This highlights the importance of optimizing both control algorithms and component selection to achieve maximum system efficiency.

The study highlights that, while advanced nonlinear control strategies provide robustness in uncertain or highly dynamic environments, simpler linear controllers may offer a better balance of performance and efficiency in well-understood systems. These conclusions can guide future implementations of DDH systems in industrial applications, promoting more sustainable and cost-effective solutions.

Author details

Juraj Benić[1*]† and Danko Brezak[2]†

1 School of Applied Mathematics and Informatics, Josip Juraj Strossmayer University of Osijek, Osijek, Croatia

2 Faculty of Mechanical Engineering and Naval Architecture, University of Zagreb, Zagreb, Croatia

† These authors contributed equally

*Address all correspondence to: jbenic@mathos.hr

References

[1] Jelali M, Kroll A. Hydraulic Servo-systems: Modelling, Identification and Control. London, London: Springer; 2003

[2] Guan C, Pan S. Adaptive sliding mode control of electro-hydraulic system with nonlinear unknown parameters. Control Engineering Practice. 2008;**16**(11):1275–1284

[3] Alleyne A, Liu R.A simplified approach to force control for electro-hydraulic systems. Control Engineering Practice. 2000;**8**(12):1347–1356

[4] Pinilla CB, Sarmiento JL, Ortiz JF. Dynamic Model and Control Design for a Nonlinear Hydraulic Actuator. In: ASME 2018 International Mechanical Engineering Congress and Exposition. Pittsburgh, Pennsylvania, USA: American Society of Mechanical Engineers; 2018. DOI: 10.1115/IMECE2018-88320

[5] Europski zakon o klimi [Internet]. https://ec.europa.eu/clima/eu-action/european-green-deal/european-climate-law_en, pristupljenostu. 2021. [Accessed: 2025-October-15]

[6] Rydberg K-E. Energy Efficient Hydraulics: System solutions for loss minimization. In: *The National Conference on Fluid Power, "Hydraulikdagar'15"*, Linköping, Sweden: 2015.

[7] Grabbel J, Ivantysynova M. An investigation of Swash plate control concepts for displacement controlled actuators. International Journal of Fluid Power. 2005;**6**(2):19–36

[8] Ivantysynova M. Innovations in Pump Design-What are future directions? In: *Proceedings of the JFPS International Symposium on Fluid Power*, Toyama, Japan: The Japan Fluid Power Systems Society; 2008.

[9] Hanawa N, Itoh M, Hirose N, Satoh H, Oda H, Hyodo T. Direct drive Volume Control of Hydraulic System and Its Application to the Ship. In: *Proceedings of the JFPS International Symposium on Fluid Power*, Tokyo, Japan: The Japan Fluid Power Systems Society; 1999.

[10] Hippalgaonkar R, Ivantysynova M. A series-parallel hydraulic hybrid mini-excavator with displacement controlled actuators. In: *13th Scandinavian International Conference on Fluid Power*, Linköping, Sweden: Linköping University Electronic Press; 2013.

[11] Imam A, Rafiq M, Jalayeri E, Sepehri N.A pump-controlled circuit for single-rod cylinders that incorporates limited throttling compensating valves. Actuators. 2018;**7**(2):13

[12] Wang H, Xu YM. The application of servo motor direct drive pump control system in pre-bending machine. Applied Mechanics and Materials. 2014;**529**:544–548.

[13] Koitto T, Kauranne H, Calonius O, Minav T, Pietola M. Experimental study on fast and energy-efficient direct driven hydraulic actuator unit. Energies. 2019;**12**(8):17

[14] Minav T, Bonato C, Sainio P, Pietola M. Direct Driven Hydraulic Drive. In: *9th International Fluid Power Conference*, Aachen, Germany: RWTH Aachen University; 2014.

[15] Minav TA, Sainio P, Pietola M. Direct-Driven Hydraulic Drive Without

Conventional Oil Tank. In: ASME/BATH 2014 Symposium on Fluid Power and Motion Control. Bath, United Kingdom: American Society of Mechanical Engineers;2014:10–12. DOI: 10.1115/FPMC2014-7834

[16] Minav TA, Heikkinen JE, Pietola M. Electric-driven zonal hydraulics in non-road mobile machinery. New Applications of Electric Drives; 2015:143–165. DOI: 10.5772/61793. [Accessed: 2015-December-09]

[17] Minav T, Papini L, Jarf A, Tammi K, Pietola M. Direct driven Hydraulics: What can possibly go wrong? -A thermal analysis. In: *XXII International Conference on Electrical Machines (ICEM)*, Lausanne, Switzerland: IEEE; 2016.

[18] Minav TA, Heikkinen JE, Pietola M. Direct driven hydraulic drive for new powertrain topologies for non-road mobile machinery. Electric Power Systems Research. 2017;**152**:390–400

[19] Niraula A, Zhang S, Minav T, Pietola M. Effect of zonal hydraulics on energy consumption and boom structure of a micro-excavator. Energies. 2018;**11**(8):2088

[20] Minav T, Heikkinen J, Schimmel T, Pietola M. Direct driven hydraulic drive: Effect of oil on efficiency in Sub-Zero conditions. Energies. 2019;**12**(2):219

[21] Minav TA, Bonato C, Sainio P, Pietola M. Efficiency of direct driven hydraulic drive for non-road mobile working machines. In: *2014 International Conference on Electrical Machines (ICEM)*, IEEE; September 2014. pp. 2431–2435

[22] Zhang S, Minav T, Pietola M. Decentralized Hydraulics for Micro Excavator. In: *15th Scandinavian International Conference on Fluid Power*, Linköping; Sweden: Linköping University Electronic Press; 2017.

[23] Agostini T, De Negri V, Minav T, Pietola M. Effect of energy recovery on efficiency in Electro-Hydrostatic closed system for differential actuator. Actuators. 2020;**9**(1):12

[24] Benić J, Brezak D, Šitum Ž, Kolar D, Lisjak D. Detailed experimental comparison of energy efficiency between proportional and direct driven hydraulic system. Results in Engineering. 2023;**18**:101178.

[25] Canudas de Wit C, Olsson H, Astrom KJ, Lischinsky P.A new model for control of systems with friction. IEEE Transactions on Automatic Control. 1995;**40**(3):419–425.

[26] Olsson H. *Control Systems with Friction*. PhD thesis, Department of Automatic Control, Lund Institute of Technology(LTH), 1996.

[27] Ruderman M. Full- and reduced-order model of hydraulic cylinder for motion control. In: *IECON 2017 - 43rd Annual Conference of the IEEE Industrial Electronics Society*, Beijing, China: IEEE; 2017.

[28] Tajjudin M, Ishak N, Ismail H, Rahiman MHF, Adnan R. Optimized PID control using Nelder-Mead method for electro-hydraulic actuator systems. In: *IEEE Control and System Graduate Research Colloquium*, Shah Alam, Malaysia: IEEE; 2011

[29] Jawaid SI, Wang L. EHA system integration design analysis and position tracking performance by PID controller. Applied Mechanics and Materials. 2013;**419**:826–834

[30] Plummer AR, Vaughan ND. Robust adaptive control for hydraulic servosystems. Journal of Dynamic Systems, Measurement, and Control. 1996;**118**(2):237–244

[31] Sirouspour MR, Salcudean SE. Nonlinear control of hydraulic robots. IEEE Transactions on Robotics and Automation. 2001;**17**(2):173–182.

[32] Kireçci A, Topalbekiroglu M, Eker I. Experimental evaluation of a model reference adaptive control for a hydraulic robot: A case study. Robotica. 2003;**21**(1):71–78

[33] Yao J, Wang C. Model reference adaptive control for a hydraulic underwater manipulator. Journal of Vibration and Control. 2011;**18**(6):893–902

[34] Sohl GA, Bobrow JE. Experiments and simulations on the nonlinear control of a hydraulic servosystem. IEEE Transactions on Control Systems Technology. 1999;7(2):238–247

[35] Köster MA. *On Modeling, Analysis and Nonlinear Control of Hydraulic Systems*. PhD thesis, Karlsruher Institut für Technologie (KIT): Fakultät für Maschinenbau; 2017.

[36] Hsieh C-T, Lai C-P. Nonlinear dynamic analysis and control of a hydraulic press electro-hydraulic servo system. Journal of Low Frequency Noise, Vibration and Active Control. 2018;**38**(3-4):1594–1606

[37] Habibi SR. Sliding mode control of a hydraulic industrial robot. Journal of Dynamic Systems, Measurement, and Control. 1999;**121**(2):312–318

[38] Jerouane M, Lamnabhi-Lagarrigue F. A new robust sliding mode controller for a hydraulic actuator. In: *Proceedings of the 40th IEEE Conference on Decision and Control (Cat. No.01CH37228)*, Orlando, FL, USA: IEEE; 2001.

[39] Wang M, Wang Y, Yang R, Fu Y, Zhu D.A sliding mode control strategy for an electrohydrostatic actuator with damping variable sliding surface. Actuators. 2020;**10**(1):3

[40] Ahn KK, Nam DNC, Jin M. Adaptive backstepping control of an electrohydraulic actuator. IEEE/ASME Transactions on Mechatronics. 2014;**19**(3):987–995

[41] Tri NM, Nam DNC, Park HG, Ahn KK. Trajectory control of an electro hydraulic actuator using an iterative backstepping control scheme. Mechatronics. 2015;**29**:96–102

[42] Vilanova R, Balaguer P. ISA-PID controller tuning: A combined min-max/ise approach. In: *2006 IEEE Conference on Computer Aided Control System Design, 2006 IEEE International Conference on Control Applications, 2006 IEEE International Symposium on Intelligent Control*, 2006. pp. 2956–2961.

[43] Pourebrahim M, Ghafari AS, Pourebrahim M. Designing a LQR controller for an electro-hydraulic-actuated-clutch model. In: *2nd International Conference on Control Science and Systems Engineering (ICCSSE)*, Singapore: IEEE; 2016.

[44] Liu B, Zhang M, Xin M, Song R, Rong X, Tian X, Tian G, Yibin L. Model independent PD-SMC method for bionic eye systems. In: *2017 Chinese Automation Congress* (CAC), 676–681, 2017.

[45] Ren Z, Zhao B, Nguyen DT. Finite-time backstepping of a nonlinear system

in strict-feedback form: Proved by Bernoulli inequality. IEEE Access. 2020;**8**:47768–47775

[46] Assegu E, Roozbahani H, Handroos H. An integrator backstepping position control of electro-hydraulic servo system based on particle swarm optimization. In: 8th FPNI Ph.D. Symposium on Fluid Power. Lappeenranta, Finland: American Society of Mechanical Engineers; 2014. DOI: 10.1115/FPNI2014-7801

[47] De Oliveira R, Guimarães Corrêa R, Kwong WH. An IMC-PID tuning procedure based on the integral square error (ise) criterion: A guide tour to understand its features. In: *IFAC Proceedings Volumes*, **28**(18):87–91, Curitiba, Brazil. September 11-12, 1995.

Chapter 3

Green Energy 4.0: An Overview of the Synergy between Machine Learning and IoT for Energy Efficiency and Emission Control

Waheeduddin Hyder, Waqas Ahmed Siddique and Mujtaba Sheikh

Abstract

Natural disasters are occurring more frequently, which emphasizes how urgent it is to cut carbon emissions and increase energy efficiency. Fossil fuel combustion for electricity generation and transportation, along with additional contributions from large-scale artificial intelligence computing and industrial activities, were the main drivers of CO_2 emissions in 2024. The integration of Industry 4.0 technologies with the Internet of Things (IoT) and machine learning (ML) into energy systems for sustainable management is known as Green Energy 4.0 (GE 4.0). IoT-based sensors enable accurate energy process monitoring by gathering real-time data on variables including demand variations, boiler and reactor conditions, and grid faults. However, manual analysis of the gathered data and taking actions may take a long time and be inaccurate. To make data analysis faster and more accurate, we can use ML. These datasets are analyzed by ML algorithms to reduce CO_2 emissions by maximizing operating efficiency, anticipating equipment breakdowns, and minimizing energy waste. By increasing thermal plant and industrial efficiency, GE 4.0 improves fossil fuel-based systems in addition to renewable ones. This chapter aims to investigate how IoT-enabled data collection and ML-driven optimization might improve energy efficiency, decrease CO_2 emissions, and facilitate the shift toward sustainable, low-carbon energy management. It will also review how IoT is used to gather data from sensors to control harmful emissions and how ML is applied to maximize the operating efficiency of energy plants.

Keywords: Green Energy 4.0, Internet of Things, machine learning, carbon emissions, energy efficiency, sustainable energy systems, renewable energy

1. Introduction

Recent natural disasters, such as floods caused by torrential rains and extreme heat waves, have emphasized the urgent need for low carbon emissions and minimizing the carbon footprint. According to the Intergovernmental Panel on Climate Change, a body of the United Nations (UN) responsible for assessing the science related to climate change, global energy-related carbon emissions were 37.8 gigatons of CO_2 [1] from carbon fuels such as coal, oil, and gas in 2024. A graph of the annual change in energy-related CO_2 emissions from 1900 to 2024 [2] shows that historically, CO_2 emissions CO_2 increased every year. However, the recent rise in industrialization and the use of electrical and electronic appliances, especially electric vehicles (EVs) and large-scale computer processing for artificial intelligence (AI), has significantly increased the demand for electrical energy and, consequently, CO_2 emissions.

Even though large-scale computer processing for AI demands a large amount of electrical generation [3, 4], we can utilize AI along with sensors to improve energy production efficiency and reduce CO_2 emissions [5, 6]. Through sensors, real-time data such as peak and off-peak demand patterns, boiler or reactor temperatures and pressures, and grid faults can be gathered to optimize efficiency, predict failures, and reduce emissions [7, 8]. This system is also known as the Internet of Things (IoT) because the data detected by the sensors are sent to the central processing facility through a widely available Internet network [9, 10]. We apply AI by using machine learning (ML) algorithms on the gathered data at the central processing facility to predict the future behavior of the system and take the necessary actions accordingly to control CO_2 emissions [11]. The integration of IoT with ML is generally termed as Green Energy 4.0 (GE 4.0). The purpose of GE 4.0 is to reduce carbon emissions by optimizing energy efficiency and enabling sustainable energy management.

Green Energy 4.0 focuses on the integration of smart, sustainable energy systems with Industry 4.0 technologies. ML plays a vital role in GE 4.0, especially when combined with the IoT [12]. Its goal is the digital transformation of the energy sector with a focus on sustainability. It combines energy sources (solar, wind, etc.) with digital technologies such as ML, IoT, and big data. Real-time insights into energy consumption, generation conditions, and operational anomalies are made possible by the IoT, which enables the continuous collection of data from a variety of sources, including smart meters, industrial sensors, renewable energy farms, and grid components. By applying ML algorithms to this data, system sustainability and dependability can be enhanced by predicting demand peaks, identifying inefficiencies, and allocating resources optimally. GE 4.0 is not limited to sustainable renewable energy, although renewable energy is its focus. It also aims to enhance energy efficiency in conventional systems to reduce energy waste in fossil fuel-based systems and optimize thermal plants and industrial processes to emit fewer carbon gases.

In light of the urgent need for sustainable energy management on a worldwide scale and the way digitalization will influence future energy systems, this chapter is driven by the opportunity to investigate how IoT and ML can contribute to the efficient implementation of Energy 4.0. Through an examination of cutting-edge methodologies and real-world implementations, this chapter seeks to illustrate how these technologies can greatly aid in reducing carbon emissions, maximizing energy efficiency, and achieving global sustainability objectives. Hereafter, we have explained the role of IoT and ML in carbon emission control by discussing the types of sensors used to monitor emissions, how IoT can be utilized to collect data from these sensors, and how ML can be applied to make intelligent decisions. Hence, we have divided this chapter into three parts, namely sensors, IoT, and ML.

2. Sensors

Let's first explore how sensors play a role in optimizing parameters for thermal treatment in power plants. The sensors are used to monitor temperature, pressure, and flow rate in thermal treatment processes.

There are many kinds of sensors used in electrical power plants to monitor the plant's operation and its output. To fully appreciate the application of IoT with ML in energy production, we should be aware of the roles and importance of these sensors in efficient power production. We must not miss the point that inefficient power production eventually leads to increased carbon emissions. Below, we have listed some sensors that are commonly used in most power plants along with their brief descriptions:

I. Electrical parameter sensors.

   i. Current transformers (CTs): Measure alternating current (AC) in high-voltage systems.

   ii. Potential transformers (PTs)/Voltage transformers (VTs): Measure high-voltage levels safely.

   iii. Power meters: Monitor active, reactive, and apparent power.

   iv. Frequency sensors: Ensure grid frequency stability (50/60 Hz).

   v. Power factor sensors: Measure the phase difference between voltage and current.

   vi. Harmonic analyzers: Detect harmonic distortions in the power system.

II. Temperature sensors

   i. Resistance temperature detectors (RTDs): High-precision temperature measurement for transformers, generators, and more.

   ii. Thermocouples: Wide-range temperature sensing (e.g., in boilers and turbines).

   iii. Infrared (IR) thermometers: Noncontact temperature monitoring of electrical components.

   iv. Thermistors: For detecting sensitive temperature changes in control systems.

III. Pressure sensors

   i. Bourdon tube sensors: Mechanical pressure measurement in pipelines.

   ii. Piezoelectric sensors: Dynamic pressure monitoring (e.g., in turbines).

   iii. Differential pressure sensors: Used in filters, cooling systems, and boilers.

   iv. Vacuum sensors: Monitor vacuum levels in condenser systems.

IV. Flow sensors

   i. Electromagnetic flow meters: Measure cooling water flow rates.

   ii. Ultrasonic flow meters: Noninvasive flow measurement in pipes.

   iii. Turbine flow meters: For fuel (gas/oil) flow monitoring.

   iv. Vortex flow meters: Detect steam or gas flow rates.

V. Vibration and speed sensors

   i. Accelerometers: Detect bearing and turbine vibrations.

   ii. Proximity sensors (inductive/capacitive): Monitors shaft displacement.

   iii. Tachometers: Measure the rotational speed of turbines and motors.

   iv. Seismic sensors: Detect abnormal vibrations in large equipment.

VI. Level sensors

   i. Float sensors: Detect water or oil levels in tanks.

   ii. Ultrasonic level sensors: Noncontact liquid level measurement.

   iii. Radar level sensors: Designed for hazardous or high-temperature environments.

   iv. Capacitive level sensors: Detect liquid or solid levels in silos.

VII. Gas and environmental sensors

   i. Oxygen (O_2) sensors: Monitor combustion efficiency in boilers.

   ii. Carbon monoxide (CO) and carbon dioxide (CO_2) sensors: Detect emissions.

iii. Hydrogen (H_2) sensors: Detect leaks in generator cooling systems.

iv. SF_6 (sulfur hexafluoride) sensors: Monitor gas-insulated switchgear.

v. Humidity sensors: Control moisture in control rooms and switchyards.

VIII. Flame and fire detection sensors

i. UV/IR flame detectors: Detect fires in boiler and fuel storage areas.

ii. Smoke detectors: Early warning for electrical fires.

iii. Heat detectors: Trigger alarms in cases of overheating.

IX. Insulation and dielectric sensors

i. Partial discharge sensors: Detect insulation breakdown in high-voltage equipment.

ii. Megohmmeters (insulation resistance testers): Check windings and cable insulation.

iii. Tan Delta (loss angle) Sensors: Assess dielectric losses in transformers.

X. Position and proximity sensors

i. Limit switches: Detect valve and circuit breaker positions.

ii. Linear variable differential transformers: Measure linear displacement.

iii. Hall effect sensors: Detect rotor position in motors.

XI. Corrosion and water quality sensors

i. pH sensors: Monitor cooling water acidity/alkalinity.

ii. Conductivity sensors: Check dissolved solids in water.

iii. Dissolved oxygen sensors: Prevent corrosion in boilers.

XII. Safety and fault detection sensors

i. Earth fault sensors: Detect ground faults in electrical systems.

ii. Arc flash detectors: Prevent electrical explosions.

iii. Surge protection sensors: Monitor lightning and switching surges.

XIII. Renewable energy-specific sensors (solar/wind)

i. Pyranometers: Measure solar irradiance.

ii. Anemometers and wind vanes: Monitor wind speed and direction.

It is beyond our scope to discuss the type of signal sensed by each sensor mentioned above and sent to the operators or feedback mechanisms, which modify or adjust the inputs according to the feedback signal for the optimum output. However, we can discuss a few of them to understand how these sensors play their role in optimizing the efficiency of electrical plants and eventually reducing carbon emissions.

2.1 Oxygen sensors

Oxygen (O_2) sensors play a vital role in the control of carbon emissions in electrical power plants that run on fossil fuels such as coal and natural gas. Maintaining the optimum air-to-fuel ratio is crucial for efficient combustion and low pollutant emissions. Oxygen itself is not considered a pollutant; however, many pollutants such as carbon monoxide (CO), nitrogen oxides (NO_x), and sulfur dioxide (SO_2) are produced during fuel combustion due to incomplete combustion. Incomplete combustion means that fuel doesn't burn completely, so instead of turning all carbon into carbon dioxide (CO_2), some of it turns into carbon monoxide (CO). Therefore, the combustion process consumes oxygen and releases pollutants in cases of incomplete combustion.

Measuring oxygen levels in exhaust or flue gas helps to estimate combustion efficiency and pollutant emissions. A low concentration of oxygen in the exhaust gas indicates incomplete combustion, which means the exhaust contains higher concentrations of pollutants such as carbon monoxide and hydrocarbons. Usually, power plants use 10–20% more oxygen than theoretically needed for complete combustion. Ideally, the supply of oxygen beyond what is required for complete combustion should not produce any pollutants. However, that is not the case due to other factors that also contribute to pollutant formation. In the case of excess oxygen, carbon monoxide (CO) may be produced due to cold engines, turbulent combustion, high flame speeds, premixed flame fronts, and quenching effects. Nitrogen oxides (NO_x) are produced when nitrogen in the air reacts with oxygen at high temperatures, which occurs even during complete combustion with excess air. Sulfur dioxide (SO_2) is produced due to sulfur in the fuel, not from incomplete combustion. Hence, pollutants are present in the exhaust gas despite complete combustion in the presence of excess oxygen. However, injecting oxygen beyond what is needed for complete combustion dilutes pollutants, meaning the oxygen spreads the pollutant molecules out, reducing the number of pollutant molecules per square inch. Nevertheless, the total mass of the pollutant (i.e., grams of CO or NO_x) per unit of time remains the same. As a result, without removing excess oxygen, measuring pollutant concentration can give a false impression of clean or pollutant-free exhaust gas.

As we mentioned earlier, 10–20% excess oxygen is used for complete combustion. However, there are different combustion setups, such as coal boilers and gas turbines, that use varying amounts of excess oxygen. To make the measurements comparable across different combustion types, emissions are recalculated or normalized to a standard oxygen level.

The emissions are “normalized” to a fixed reference O_2 level (e.g., 3% or 6%, depending on the fuel type). This eliminates the effect of dilution and reflects the true emission level per unit of exhaust.

According to Directive 2010/75/EU [13], the type of fuel or combustion technology determines the reference O_2 level, as given in **Table 1**.

Let’s understand normalization with a simple example. Suppose we have three power plants.

1. **Plant A**: Coal-fired (solid fuel)

2. **Plant B:** Gas-fired (gaseous fuel)

3. **Plant C:** Gas turbine (also gas, but uses a different technology)

The measured NOx and O_2 emissions are shown in **Table 2**.

For normalization we have:

$$E_N = E_M \frac{21 - O_2^{ref}}{21 - O_2^M} \tag{1}$$

where

E_N – normalized emission

E_M – measured emission

O_2^{ref} – reference oxygen, that is, 6%

O_2^M – measured oxygen

21 – percentage of oxygen in air.

Now, we calculate the normalization values for each power plant.

S. No.	Type of fuel/technology	Reference O_2 level
1	Solid fuels (coal, biomass)	6%
2	Liquid fuels (oil)	3%
3	Gaseous fuels (natural gas)	3%
4	Gas turbines	15%
5	Waste incineration	11%

Table 1.
Oxygen levels for different types of fuels.

Plant	Fuel type	Measured NO_x (mg/Nm^3)	Measured O_2 (%) in exhaust
A	Coal (solid)	180	5%
B	Natural gas	120	4%
C	Gas turbine (NG)	90	12%

Table 2.
Nitrogen oxide and oxygen levels for Plants A, B, and C.

Normalized emission for Plant A is:

$$E_N^A = E_M \frac{21 - O_2^{ref}}{21 - O_2^M} \tag{2}$$

$$E_N^A = 180 \frac{21 - 6}{21 - 9} = 187.50\, mg/Nm^3 \tag{3}$$

Normalized emission for Plant B is:

$$E_N^A = E_M \frac{21 - O_2^{ref}}{21 - O_2^M} \tag{4}$$

$$E_N^A = 120 \frac{21 - 3}{21 - 4} = 127.06\, mg/Nm^3 \tag{5}$$

Normalized emission for Plant C is:

$$E_N^A = E_M \frac{21 - O_2^{ref}}{21 - O_2^M} \tag{6}$$

$$E_N^A = 90 \frac{21 - 15}{21 - 12} = 60.00\, mg/Nm^3 \tag{7}$$

From the normalization results, we find that Plant B's NO_x appears lower than Plant A's, but it is actually higher than expected. Additionally, we learned that Plant C has the cleanest performance, and even though the dilution is high (12% O_2), its performance proves to be the best when normalized to 15%.

Similarly, we can normalize the outputs when the type of fuel is the same to adjust the readings to the same reference point (6%). Let's assume that we have two power plants, A and B, that use coal combustion and emit NO_x. Assume that Plant A consumes less excess oxygen compared to Plant B, as shown in **Table 3**.

The values of NO_x in **Table 3** show that Plant B is emitting less pollutant flue gas, but that is misleading. According to EU regulation for coal (Directive 2010/75/EU on Industrial Emissions [13]), emissions must be reported at 6% O_2. For that, we normalize NO_x to a standard 6% O_2 by the following:

Normalized emission for Plant A is:

$$E_N^A = E_M \frac{21 - O_2^{ref}}{21 - O_2^M} \tag{8}$$

Plant	Measured NO_x (mg/Nm^3)	Measured O_2 (%) in exhaust
A	180 mg/Nm^3	5% O_2
B	150 mg/Nm^3	9% O_2

Table 3.
Nitrogen oxide and oxygen levels for Plants A and B.

$$E_N^A = 180\frac{21-6}{21-9} = 187.5\,mg/Nm^3 \tag{9}$$

Normalized emission for Plant B is:

$$E_N^B = E_M\frac{21-O_2^{ref}}{21-O_2^M} \tag{10}$$

$$E_N^A = 150\frac{21-6}{21-5} = 168.75\,mg/Nm^3 \tag{11}$$

Hence, after normalization, we find that **Plant A** is polluting more than **Plant B**.

2.2 Turbine flow meter

There are some other sensors that do not play any direct role in the control of CO_2 emissions but have an important part in CO_2 emission control indirectly, such as a **turbine flow meter**. Turbine flow meters measure the volume or mass flow rate of fuels such as natural gas and oil that are used in the combustion system. This helps to monitor and adjust the flow of gas to avoid excess gas injections in the combustion system.

2.3 Zirconium dioxide sensor

The zirconium dioxide sensor is commonly used for oxygen measurements in flue gas, and it is installed in the flue gas duct. It works based on the difference in oxygen concentration between flue gas and reference air. It produces a voltage that correlates with oxygen concentration. The voltage value is converted to a digital signal using an analog-to-digital converter. Data are sent via protocols such as MQTT, HTTP, and Modbus over TCP/IP (we will discuss these protocols in the IoT section).

2.4 Pyranometer

Solar panels use irradiance sensors (also called pyranometers) to measure sunlight intensity. These sensors are important devices in solar energy systems for measuring the amount of solar radiation reaching the surface. Irradiance is measured in W/m^2. They are used to measure irradiance at different angles to optimize the horizontal and vertical positioning of solar panels using sun-tracking systems. The optimum angles of solar panels, where they receive the maximum radiation from the sun and produce the maximum output, vary due to factors such as clouds, dust, and seasonal changes.

2.5 Anemometers

Anemometers measure wind speed to optimize turbine performance and prevent damage. Traditional anemometers use a cup-like device to catch the wind, rotate the wheel, and measure the speed of the wind (see **Figure 1**).

Wind affects the speed of sound in air, so the travel time of sound changes depending on wind direction and speed. This can be used to measure wind speed

Figure 1.
An anemometer used for measuring wind speed and direction for wind turbine performance monitoring [14].

with ultrasonic anemometers. Ultrasonic anemometers calculate wind velocity by measuring the time it takes for ultrasonic pulses to travel between pairs of ultrasonic transmitters and receivers (see **Figure 2**).

2.6 Wind vanes

Wind turbines also use sensors to control the speed of the turbine, the direction of the wind to control yaw, and vibrations. Mechanical wind vanes determine the wind direction and allow the turbine to adjust its position for optimal energy capture. A typical wind vane has an arrow that rotates freely on a vertical axis. The tail catches the wind, causing the vane to rotate until it points in the direction of the wind. Electronic components, such as potentiometers, convert the mechanical rotation into an electrical signal. The yaw system of the turbine rotates the nacelle to keep the blades perpendicular to the wind (see **Figure 3**).

2.7 Speed monitoring sensors

To monitor the rotational speed, wind turbines use revolutions per minute (RPM) sensors. These sensors ensure safe operation and prevent overspeed conditions. Hall effect sensors are commonly used for rotor and gearbox speed monitoring. Optical encoders are frequently employed in pitch systems and generator monitoring.

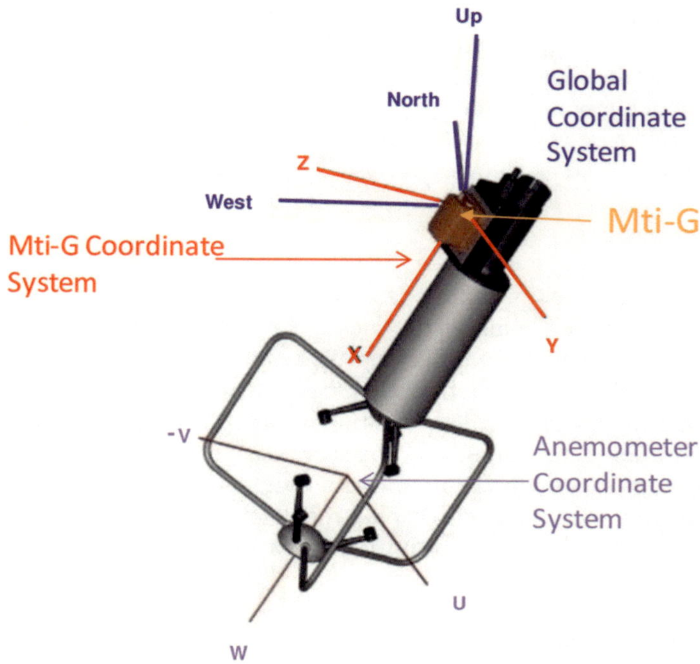

Figure 2.
Ultrasonic anemometer used for measuring wind speed and direction by analyzing the time-of-flight of ultrasonic pulses, offering high accuracy and reliability [15].

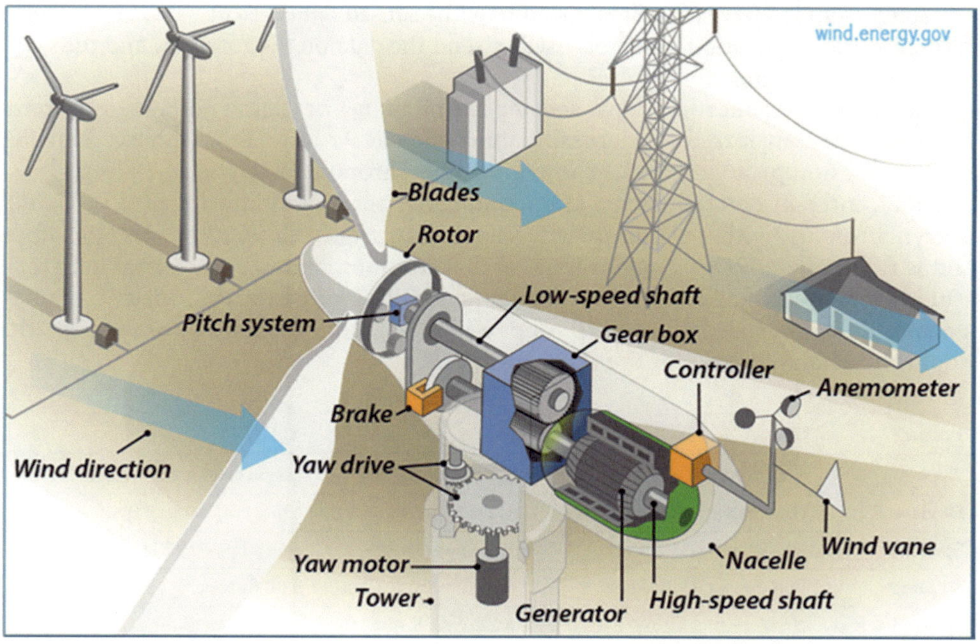

Figure 3.
Components of a typical wind turbine (Source: U.S. Department of Energy) [16].

Inductive (proximity) sensors are utilized in gearbox and shaft speed monitoring. If the rotational speed of the turbine exceeds the threshold value, the braking system is applied to reduce the speed. The braking system uses a pitch control system to rotate the blades, reducing wind contact and slowing down the rotor. Electrical brakes, such as generator torque control, are applied in cases of high winds or faults. Disc brakes can also be used as an emergency backup if pitch control fails.

2.8 Vibration sensors

Vibration sensors are critical for detecting faults such as imbalances caused by wear and tear in rotating components such as bearings and gearboxes. Piezoelectric accelerometers are the most common type, where a crystal (e.g., quartz or ceramic) generates an electric charge when mechanically stressed. A seismic mass is attached to the crystal, compressing it when vibration occurs. These sensors are used for high-frequency vibration detection in gearboxes or bearings. Velocity sensors are useful for mid-frequency range vibrations and are often employed for structural health monitoring (e.g., tower, nacelle). Displacement sensors are used for low-frequency vibration monitoring, such as rotor imbalance or misalignment.

3. Internet of Things (IoT)

Internet of Things (IoT) is a broad concept that includes diverse devices such as sensors, actuators, smartphones, and cloud systems connected to the Internet via communication protocols. The gathered data are sent to a data collection point, which may be a data center on the cloud where the data is processed and information is extracted from it. Here, we will focus on how the sensed data are converted into packet form, the communication protocols used to send the data on the Internet, and the network architecture.

The signals produced by most sensors are in analog or digital voltage or current form and are transferred to processing units such as Arduino or Raspberry Pi using communication protocols such as universal asynchronous receiver/transmitter (UART), inter-integrated circuit (I^2C), and serial peripheral interface (SPI). UART is asynchronous, with a communication speed ranging from 9600 bps to 115.2 kbps and is full duplex; I2C is synchronous, with a communication speed ranging from 100 kbps to 3,400 kbps and is half duplex; and SPI is synchronous, with a communication speed of up to 10 Mbps and is full duplex. **Table 4** summarizes the important characteristics of the three protocols.

Feature	SPI	I2C	UART
Type	Synchronous	Synchronous	Asynchronous
Devices	One master, multiple slaves	Multidevice bus	Only two devices (point-to-point)
Speed	10 + Mbps	100 kbps–3.4 Mbps	9600 bps–115.2 kbps
Duplex	Full duplex	Half duplex	Full duplex
Error check	None	ACK/NACK bits	None

Table 4.
Comparison of common serial communication protocols – SPI, I^2C, and UART – based on type, device support, speed, duplex capability, and error-checking.

The data received by microcontrollers such as Arduino, Raspberry Pi, ESP32 (Espressif Systems), STM32 (STMicroelectronics), C2000 (Texas Instruments), NXP Kinetis, PIC32 (Microchip), and Renesas RX/RA series are formatted into structured data such as JSON (JavaScript Object Notation), CSV (Comma-Separated Values), or custom binary protocols for processing, storage, or transmission.

JSON is commonly used in Wi-Fi devices such as ESP32 whereas CSV is commonly used for logging in Arduino or STM32. Binary protocols, for example, Modbus, are used in industrial microcontrollers such as C2000 or PIC32.

The structured data are sent to the cloud via IoT protocols. Some popular protocols at different layers are shown in **Table 5**.

3.1 Physical layer protocols

LoRa (Long Range) is a low-power, long-range wireless communication protocol (15–20 km in an open field) mainly used in IoT applications. It is a modulation technique developed by Semtech Corporation, based on chirp spread spectrum (CSS) technology, with low power consumption, making it suitable for battery-operated IoT devices. It uses Advanced Encryption Standard (AES) encryption for data security, which makes it reliable and dependable for Industrial IoT and environmental monitoring.

LoRa is supported by most of the microcontrollers mentioned above. The output of the microcontroller is sent to the LoRa module, which handles wireless communication. For example, the SX1276/78 is used with Arduino. However, there are some microcontrollers that have built-in LoRa support. A list of some LoRa modules with their features is provided in **Table 6**.

In addition to LoRa, Wi-Fi, Zigbee, and BLE are also used for connectivity between the IoT device and the IoT gateway.

3.2 Network layer protocols

3.2.1 IPV6

To route packets from the sensor to the IoT gateway, network protocols such as IPv6, 6LoWPAN, Thread, and RPL are commonly used. IPv6 is a standard Internet protocol developed by the Internet Engineering Task Force (IETF) in 1995 as RFC 2460, later updated by RFC 2800 in 2017. The main advantage of IPv6 is that it uses a 128-bit address, which can produce approximately 340

Protocol layer	**Examples**
Physical/link	LoRa, Wi-Fi, Zigbee, Bluetooth Low Energy (BLE)
Network	IPv6, 6LoWPAN, Thread, RPL
Transport	UDP, TCP
Application (IoT)	MQTT, CoAP, HTTP/REST, AMQP

Table 5.
Commonly used protocols in IoT.

LoRa module	Features	Notes
Ra-01 (SX1278)	433 MHz, SPI, small module	Used with Arduino/ESP
RFM95/96/98	868/915 MHz, SPI	Used with STM32
Heltec WiFi LoRa 32	ESP32 + OLED + LoRa (SX1276)	All-in-one, easy prototyping
TTGO LoRa32	ESP32 + OLED + SX1278	low cost
STM32WL55	STM32 + built-in LoRa	No external module needed

Table 6.
Ra-01 (SX1278) by Ai-Thinker, based on the Semtech SX1278 chip, supports the 433 MHz band. RFM95/96/98, a family of HopeRF LoRa transceiver modules built on Semtech SX1276/77/78/79 chips, supports the 433/868/915 MHz bands. Heltec WiFi LoRa 32 is an all-in-one IoT development board ESP32 microcontroller and an OLED display. TTGO LoRa32, a development board by LilyGO, includes ESP32 microcontroller and an OLED display. STM32WL55 is a system-on-chip from STMicroelectronics with a built-in LoRa transceiver.

undecillion (3.4 x 10^{38}) unique addresses. The packet processing of IPv6 is more efficient than that of IPv4.

3.2.2 6LoWPAN

6LoWPAN is an adaptation of IPv6 for low-power, low-bandwidth networks such as IEEE 802.15.4. 6LoWPAN stands for *IPv6 over Low-power Wireless Personal Area Networks*. It provides compression, fragmentation, and reassembly to make IPv6 feasible on constrained networks. It facilitates end-to-end IP connectivity for small-frame wireless networks, making it suitable for IoT applications. It reduces the large IPv6 header (40 bytes) to just a few bytes (2–7 bytes) using context-aware compression. Compression is performed using IP header compression as defined in RFC 6282. Certain fields in IPv6 headers often have default values and can be omitted. For example, the version field is always 6 and can be omitted. Similarly, the traffic class and flow label are often 0 and can also be omitted. If devices are on the same local network (link-local), 6LoWPAN can omit the prefix (e.g., FE80::).

3.2.3 Thread

Thread is also an IPv6-based mesh networking protocol designed specifically for smart networks and IoT devices. It enables reliable, secure, and scalable communication between low-power devices. Mesh networking is used to reduce a single point of failure by connecting routers with each other and forming a self-healing mesh network. A self-healing mesh network automatically detects broken communication paths and reroutes traffic through alternate paths. Routers discover each other when powered on and send out discovery messages (multicast) over IEEE 802.15.4. Nearby routers respond, and the device joins the mesh by establishing secure links. Each Thread router maintains a neighbor table of adjacent routers using the Mesh Link Establishment protocol. One router becomes the leader through a dynamic election process. If the leader goes offline, another router is automatically promoted to leader. If a router goes offline or a link degrades, neighboring routers remove it from their tables, and packets are rerouted through other routers.

3.2.4 Routing Protocol for Low-Power and Lossy Networks (RPL)

RPL is also designed by the IETF for IPv6-based mesh networks for IoT sensors and actuators. Lossy networks have unstable connections due to weak signals in factories or homes because of obstacles or other noises. Like Thread, it is also a mesh-based network and self-healing.

3.3 Transport layer protocols

On the transport layer, the User Datagram Protocol (UDP) or Transmission Control Protocol (TCP) is applied. UDP is more commonly used in IoT stacks due to its simplicity, low latency, and low power consumption. TCP is utilized where reliability and ordered delivery of packets are critical.

3.4 Application layer protocols

3.4.1 MQTT

On the application layer, Message Queuing Telemetry Transport (MQTT), Constrained Application Protocol (CoAP), Hypertext Transfer Protocol (HTTP), Hypertext Transfer Protocol Secure (HTTPS), Advanced Message Queuing Protocol (AMQP), Extensible Messaging and Presence Protocol (XMPP), and Data Distribution Service (DDS) are used.

MQTT is the most popular IoT protocol, and it is designed for low-bandwidth, high-latency, and unreliable networks. MQTT formats messages using a two-byte header and a payload. MQTT uses a publish–subscribe model for communication. A device either wants to send data or receive data from another device. If it sends data, it is called publishing, and if it requests data from another device, it is called subscribing. In MQTT, the term *topic* refers to a string that acts as an address or label. Topics are used to organize and route messages between publishers and subscribers. Publishers and subscribers send and receive data via a *broker,* which serves as a central server for communication between them. A device sends its data to the broker whenever it has data to publish, and subscribers retrieve the data of interest from the broker if it is available. Publishers and subscribers do not communicate with each other directly. For example, a temperature sensor (publisher) sends temperature data every 5 seconds to the topic *home/temperature*. Even if there is no subscriber present, the publisher still publishes the message to the broker. If no one is subscribed, the message may be discarded. Popular MQTT brokers are listed in **Table 7**.

Broker name	Description
Mosquitto	Lightweight, open-source MQTT broker
EMQX	High-performance, scalable broker
HiveMQ	Enterprise-level broker with Web UI
VerneMQ	MQTT broker designed for high availability

Table 7.
Popular MQTT brokers commonly used in IoT systems, including open-source and cloud-based platforms that provide lightweight publish/subscribe communication for scalable device connectivity.

3.4.2 CoAP

Like MQTT, CoAP is also designed for resource-constrained devices and low-power, lossy networks. It is standardized by the IETF Constrained RESTful Environments (CoRE) working group under RFC 7252. CoAP is based on a client-server model, just like HTTP. A client sends a GET request to a resource on a server (e.g., a sensor), and the server replies with the sensor value. The CoAP packet format is given in **Figure 4.** [17].

Ver: 2 bits – CoAP version. Always 01.

T: 2 bits – Defines the transaction type. Indicates whether this message is of type Confirmable (0), Nonconfirmable (1), Acknowledgment (2), or Reset (3).

TKL: 4 bits – Indicates the length of the variable-length Token field (0–8 bytes).

Code: Request or response code – 8 bits

Message ID: 16 bits

Token: 32 bits

Options: 32 bits

Payload Marker: 8 bits

Payload: 24 bits

The CoAP Method Codes listed in RFC 7252, Section 5.9.1, use an 8-bit (1-byte) field where the 3 MSBs are 0 to indicate that it is a *request,* and the type of request is indicated by the 5 LSBs (**Table 8**).

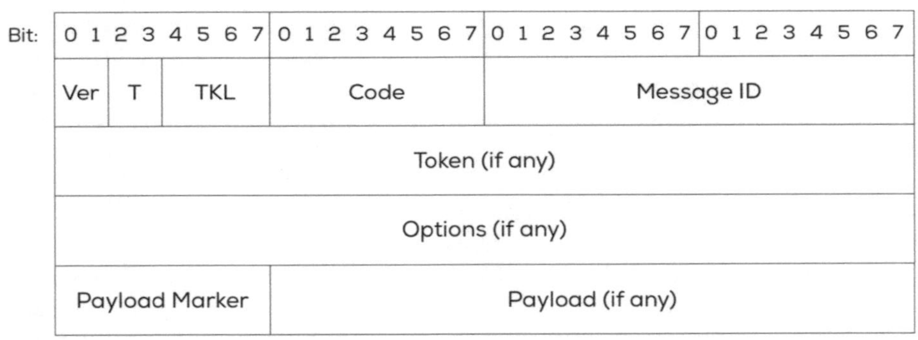

Figure 4.
CoAP packet format illustrating the header structure, including version, type, token length, code, message ID, options, and payload, used for lightweight communication in IoT applications.

Type	Code	Binary (8 bits)	
GET	0.01	000	0,0001
POST	0.02	000	0,0010
PUT	0.03	000	00011
DELETE	0.04	000	0,0100

Table 8.
CoAP request codes bit division.

Similarly, to indicate the message is a response, the 3 MSBs show the response class (2 for success, 4 for client error, and 5 for server error), and the 5 LSBs indicate the response detail (**Table 9**).

Suppose there is a temperature sensor in a smart home that acts as a CoAP server, and a mobile application that acts as a CoAP client. When the mobile application wants to get the current temperature from the sensor, it sends a GET request to the resource "/temperature" on the CoAP server.

3.4.3 HTTP/HTTPS

Since HTTP and HTTPS are widely used protocols for web applications, a web application like a web browser sends a request to download a web page from the server (e.g., GET /index.html). The server processes the request and sends back the requested web page to the web browser. The request methods of HTTP are similar to those of CoAP, that is, GET, POST, PUT, DELETE, and HEAD. However, the data are sent in plain text from the server to the client, which poses security risks such as eavesdropping (sniffing), man-in-the-middle attacks, and session hijacking, where cookies can be stolen. To make HTTP secure, data are encrypted using the AES, and the connection is encrypted using Secure Sockets Layer or Transport Layer Security in HTTPS. Today, HTTPS is more widely used than HTTP because of its enhanced security features.

3.4.4 AMQP

AMQP is a messaging protocol used for sending and receiving messages in IoT, ensuring reliable and secure communication. Like MQTT, AMQP also uses a *broker* to manage communication between the sender and the receiver. The received messages on the server are *queued* to hold messages until they are consumed. *Exchanges* route

Type	Code	Binary (8 bits)	
CREATED	2.01	010	00001
DELETED	2.02	010	00010
VALID	2.03	010	00011
CHANGED	2.04	010	00,100
CONTENT	2.05	010	00101
BAD REQUEST	4.00	100	00000
UNAUTHORIZED	4.01	100	00001
NOT FOUND	4.04	100	00100
METHOD NOT ALLOWED	4.05	100	00101
INTERNAL SERVER ERROR	5.00	101	00000
SERVICE UNAVAIALABLE	5.03	101	00011

Table 9.
CoAP response codes bit division.

messages to the correct queue using *bindings,* which define how messages are routed from exchanges to queues.

Suppose a sensor device publishes a message (e.g., *{"temp":25}*) to be sent to a server known as a broker. The exchange routes the message to the appropriate queue based on the following:

Direct exchange (e.g., sen.temp)

Topic exchange (e.g., sen.*)

Fanout exchange (e.g., broadcast to all queues)

A backend service, for example, a cloud server, subscribes to the queue and stores the data in a database or generates alerts.

3.4.5 XMPP

XMPP is a machine-to-machine-based communication protocol that provides decentralized and secure messaging. XMPP allows multiple devices to communicate with different servers. Each device has a unique identification, similar to email address (sen1@mynet.iot.net), called a Jabber ID (JID). Like MQTT, AMQP also works on a publish–subscribe model.

More functions can be added using XMPP Extension Protocols (XEPs) for IoT-specific needs:

1. XEP-0323 (IoT sensor data): Standardizes the exchange of sensor data.

2. XEP-0347 (IoT control): Enables remote device control.

3. XEP-0325 (IoT provisioning): Manages device registration.

4. XEP-0384 (Off-the-Record Messaging, Extended and Modernized encryption): End-to-end encrypted messaging.

5. XEP-0198 (Stream management): Resumes broken connections, which is useful for IoT.

6. XEP-0363 (HTTP file upload): Useful for firmware updates.

3.4.6 DDS

DDS is a decentralized model that operates on a peer-to-peer architecture. Devices communicate directly with each other without relying on any server or broker. This eliminates the risk of a single point of failure. It was designed for low-latency and high-throughput communication, which makes it suitable for real-time, critical IoT applications. Devices are connected to each other through automatic discovery methods. DDS is a data-centric protocol that focuses on data rather than devices. The sender (publisher) and receiver (subscriber) exchange data based on topics (data types). DDS is particularly useful for real-time monitoring of factory equipment, traffic management, and vehicle-to-vehicle communication.

4. Machine learning

Now, we briefly describe ML and its methodologies, and then we analyze the implementation of ML in IoT. ML is a subdomain of AI that uses statistical methods to estimate future values. ML methods can be categorized into three classes, namely supervised learning, unsupervised learning, and reinforcement learning. Linear or polynomial estimation equations are used to predict future values. The most common linear equation used for ML is $y = mx + b$, where y is the output, x is the input, m is the gradient, and b is the y-intercept.

Suppose we have data on car sales for a year, as shown in **Figure 5**. Our objective is to find a line that passes through the data points in such a way that the vertical distance between each data point and the line is minimized. This is called the best-fit line or regression, which is shown in the graph as a continuous line A. The other dashed lines also pass through the data points, but their vertical distance from all the data points is not minimal. The vertical distance between a data point and the line is called the error or residual.

To find the best-fit line, we need the gradient m and y-intercept b values based on the input values of x. In **Figure 5,** the continuous line is the best-fit line because it has the minimum error or vertical distance with all the dataset points. The values of m and b are called the trained weights or parameters that are used to predict y based on input features x. This is called the linear regression model. Linear regression is the most common type of supervised ML algorithm used to predict time-series-based activities. It is called a supervised ML model because, for each set of features, the output y is known, and based on that trained model, the output is predicted.

Compared to supervised learning, where the training data are labeled, unsupervised learning groups the data during training. The data are grouped based on its feature values. Features with similar values are considered part of a group. An example of unsupervised learning grouping is shown in **Figure 6**.

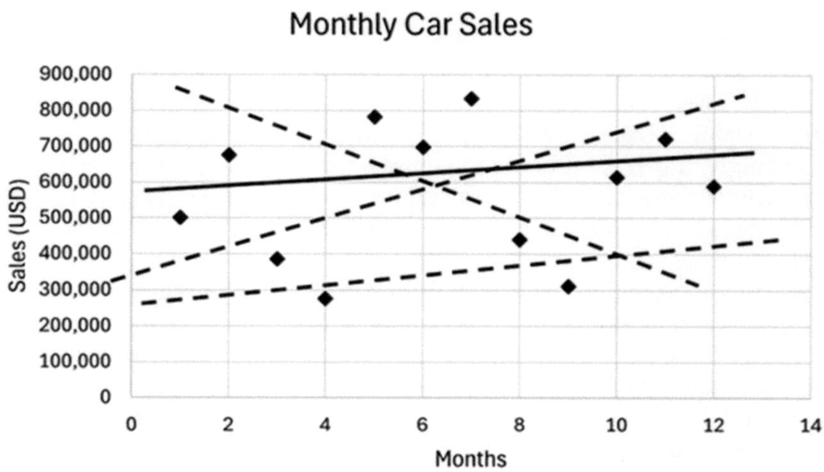

Figure 5.
Linear regression: monthly car sales trend analysis showing sales values (USD) across different months, with regression line (solid). Continuous line is the best fit line.

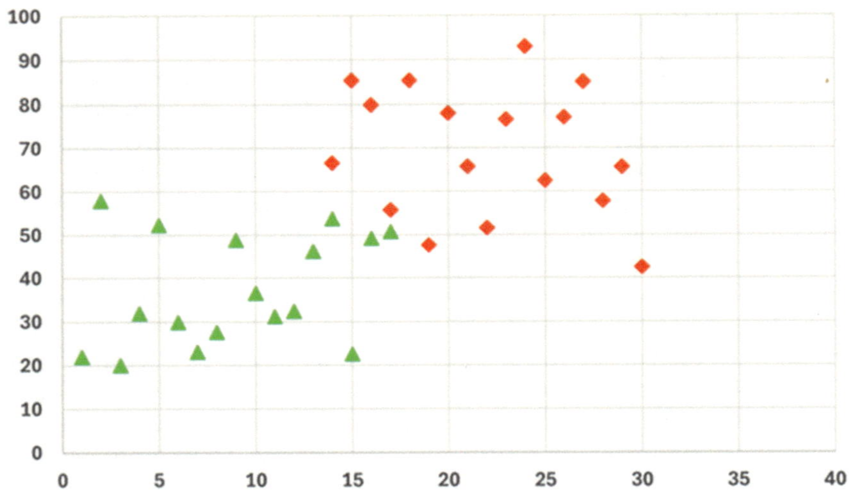

Figure 6.
Clustered scatter plot showing two distinct data groups, represented by green triangles and red diamonds, illustrating class separation and potential clustering patterns in unsupervised learning.

Reinforcement learning is entirely different from supervised learning and unsupervised learning methods in the sense that it does not rely on training data. The agent starts with random actions according to the given rules and policies and measures the feedback in the form of rewards or penalties against predefined goals. As the learning progresses, the actions taken by the agent become more well-informed rather than random.

5. IoT with ML

Now we see how learning algorithms can be combined with IoT to increase the efficiency of systems. We begin with an example of a solar system. Irradiance sensors, when integrated with IoT and ML, enable efficient solar energy forecasting and optimization. Using ML, we can predict future power production and adjust the load accordingly. Supervised learning algorithms such as long short-term memory (LSTM), Prophet, and ARIMA are trained on labeled data. Underperforming panels can also be detected by ML algorithms such as Isolation Forest, Autoencoders, and SVM.

The performance of photovoltaic cells in solar panels degrades over time, similar to other electrical and electronic components. A typical degradation value is approximately 0.5% per °C above 25°C [18]. Overheating and thermal stress may cause hairline fractures in silicon solar cells (often invisible to the naked eye), delamination, and solder fatigue, that is, cracking of solder connections between solar cells in solar panels. If the temperature rise is too high, it may result in permanent damage to the solar panels. To prevent overheating, cooling methods are applied. There are several active cooling methods, such as fans, water sprinklers, circulating coolants in pipes behind the solar panels, and solid-state heat pumps. Temperatures are measured using different types of sensors, including thermocouples, RTDs, thermistors, infrared (IR) sensors, and semiconductor-based sensors, for example, DS18B20 or DHT22. RTDs and thermistors may be attached to panels or placed in the air, whereas IR sensors measure surface temperature without physical contact. Sensors send temperature data to microcontrollers or IoT gateways. Microcracks can be detected using electroluminescence imaging, which identifies cracks as dark lines, or IR thermography. These techniques require manual inspection or can be performed using drones. The images obtained can be analyzed by convolutional neural networks to identify microcracks in the solar panels. Delamination faults can be detected using ultrasonic testing (manual or robotic) to check for air gaps and impedance spectroscopy (embedded sensors) to measure moisture ingress.

These sensors send data to an IoT controller to detect and predict any faults. The pattern of data collected from these sensors is used to predict failures or issues in advance using ML algorithms. Supervised ML algorithms may be used to predict energy production of solar or wind turbines, load requirements, maintenance needs, and faults. Supervised learning algorithms are more commonly used when labeled data are available from IoT sensors, for example, energy readings and device states. When we intend to determine energy usage patterns or anomalies and the data are not labeled, unsupervised learning algorithms are used. For example, solar panels have many sensors that constantly measure variables such as temperature, voltage, current, and irradiance. These sensors generate a large amount of data every day. However, not all of this data are labeled. We only know at what temperature overheating occurred or at what irradiance magnitude we get the maximum output. Hence, the algorithm learns patterns of faults from the small labeled dataset where we know a fault happened. Then, it uses unlabeled data from the sensors to improve model learning. A list of some supervised, unsupervised, and semisupervised learning algorithms and their possible applications is given in **Table 10.**

Machine learning can play a vital role in enhancing CO_2 control in combustion systems by optimizing combustion efficiency and reducing emissions. The amount of air in the fuel mixture, its temperature, and pressure have a significant effect on combustion efficiency and consequently on CO_2 emissions. For example, high flame temperatures in gas turbines cause an increase in NO_x emissions, whereas low flame temperatures cause an increase in CO_2 emissions. IR pyrometers or thermocouples send data to processing units, where ML algorithms such as random forest, XGBoost, and LSTM may be used to predict optimal flame temperatures or forecast flame instability trends, respectively.

5.1 Linear logistic regression

Now, let's describe some more commonly used ML algorithms applied to IoT for green energy. We have already described linear regression in Section 4. Linear logistic regression is used for the binary classification of data based on their features. It involves finding a straight line that separates the two data types based on their features. For example, suppose a smart factory monitors CO_2 emissions in real time to predict whether CO_2 levels exceed a dangerous threshold. Logistic regression can be used to predict the CO_2 emission level. Typically, 5,000 to 10,000 samples of CO_2 emissions are sufficient; however, a higher number of samples may be required if the environment is more complex. Sample data are shown in **Table 11**.

Algorithm	Use case in Green Energy IoT
	Supervised learning
Linear regression	Predict energy demand and supply.
Logistic regression	Classify equipment status (faulty/healthy).
Decision trees	Energy demand response and load forecasting.
Random forests	Predictive maintenance using sensor data.
Support vector machines (SVMs)	Fault detection and load forecasting in solar panels or wind turbines.
K-nearest neighbors (KNN)	Classify energy usage patterns.
Gradient boosting (e.g., XGBoost, LightGBM)	Forecast energy prices and optimize grid operations.
Neural networks (MLP, ANN)	Time series forecasting for power generation and consumption.
	Unsupervised learning
K-means clustering	Segment smart meter data or group users based on consumption.
DBSCAN/HDBSCAN	Detect anomalies in energy data from IoT sensors.
Principal component analysis (PCA)	Reduce the dimensionality of multivariate IoT sensor data.
Autoencoders	Anomaly detection in grid equipment.
	Semisupervised learning
Label propagation	Improve the classification of sensor events with limited labels.
Self-training neural networks	Predict grid failures using partially labeled data.

Table 10.
Supervised, unsupervised, and semisupervised learning algorithms and their possible applications.

CO_2 level (ppm)	Temp (°C)	Humidity (%)	RPM	Alert (1/0)
850	26	40	500	0
1200	34	65	900	1
780	25	38	450	0
1050	31	55	750	1
920	29	48	600	0

Table 11.
Sample dataset from a smart factory monitoring system with corresponding alert signals, used for training ML models.

One might wonder how it is different from conventional alert systems designed using simple electronics or microcontrollers. In conventional systems, a sensor may continuously sense the condition and send signals to the electronic control system, which triggers an alert signal whenever CO_2 level exceeds 1000 ppm. Typically, electronic control systems are designed to generate alerts for a single feature, such CO_2 level, or for two or three additional features and a few combinations of them. However, ML can work with as many features as one may desire and numerous possible combinations by training the model using past data.

5.2 Decision tree

A decision tree is also a supervised learning algorithm used for classification and regression. It creates a tree-like structure by splitting data into subsets based on feature values. There are nodes, branches, and leaves in a decision tree. Nodes are based on decision rules such as "Is solar radiance >3 kWh/sqm/day," branches represent decision outcomes, and leaves indicate predictions. Suppose we have a solar system that uses intensity, panel temperature, battery charge level, and energy demand to decide whether to supply more energy to the grid, charge batteries, supply to the load, or use a backup source. A sample decision tree for the given scenario is shown in **Figure 7**.

Table 12 shows sample data collection and decisions based on the decision tree as shown in **Figure 7**.

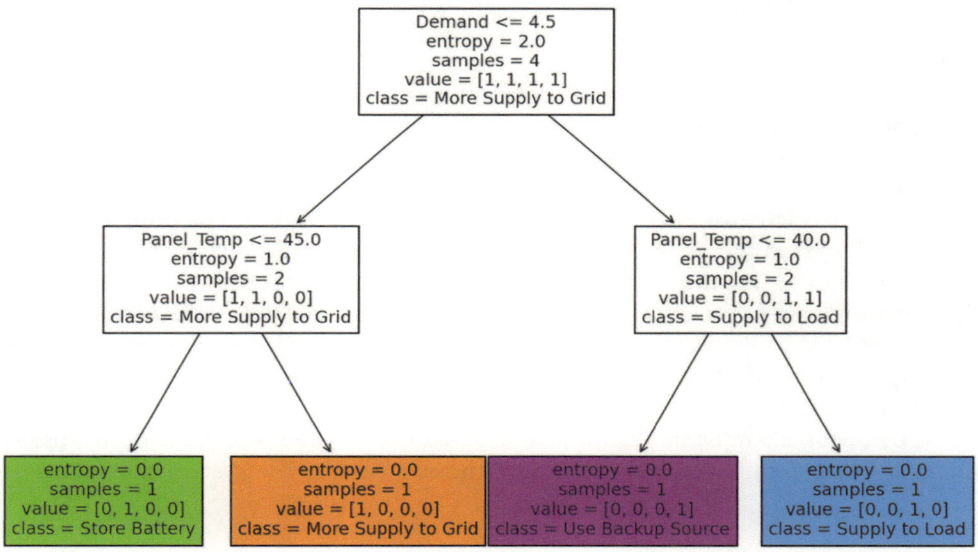

Figure 7.
Decision tree model for smart energy management, illustrating how demand and panel temperature influence decisions on whether to store energy in the battery, supply energy to the grid, supply to the load, or use a backup source.

Intensity (Lux)	Panel temp (°C)	Battery (%)	Demand (kW)	Decision
10000	35	90%	5	Use backup source
30000	45	50%	7	Supply to load
50000	50	100%	1	More supply to grid
60000	40	20%	4	Store battery

Table 12.
*A sample data collection and decisions based on the decision tree as shown in **Figure 7**.*

5.3 Random forest

The decision tree algorithm is prone to overfitting, that is, it predicts well for the trained dataset but fails to predict for new values that are different from the trained dataset. Random forests use multiple decision trees and combine their results to overcome the problem of overfitting. From the original dataset, samples are selected randomly with replacement to create multiple subsets, and each subset is used to train a decision tree. This is called bootstrap sampling. To further reduce overfitting, only a random subset of features is considered instead of all features. This reduces correlation between trees and improves generalization. When the results are combined, majority voting from all trees is used for classification, and the average of predictions from all trees is used for regression. Suppose an IoT system collects data from a wind turbine farm to monitor its performance. At intervals of every few seconds, sensors collect wind speed (m/s), blade angle (degrees), and rotor RPM to predict the power output based on the sensor readings. Random forest is used to predict power output using the sensor readings.

5.4 Support vector machine

Support vector machine is a supervised learning algorithm used mainly for classification, and it finds the best boundary, also called a hyperplane, that separates data points of different classes with the largest margin. The decision boundary is called a hyperplane because its dimensions change according to the number of features. For example, for two features, the dimension is 1D, that is, a straight line; for three features, the dimension of the plane is 2D, and so on. Hence, we can say that for *n* features, we have a hyperplane of n-1 dimensions. The data points closest to the hyperplane are called support vectors. We aim to have the largest distance between the hyperplane and the nearest data point. This distance is called the margin, as shown in **Figure 8**.

Let's see an example of a combustion-based power generator with an IoT-based monitoring system to continuously track emission levels. The system should

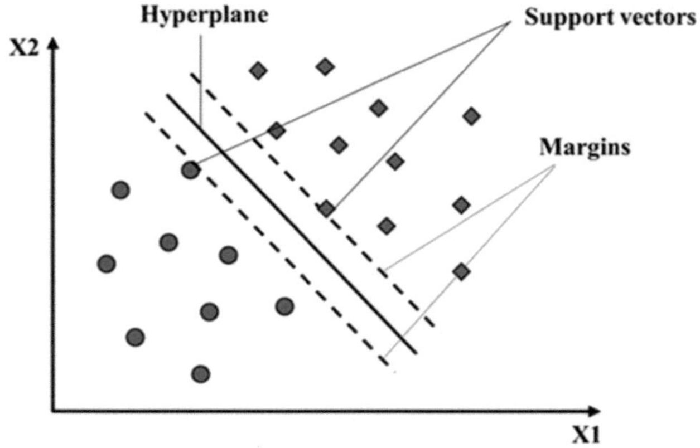

Figure 8.
SVM classification illustrates the optimal hyperplane that separates two classes with maximum margin, supported by data points known as support vectors.

automatically decide when to adjust the combustion process (air-to-fuel ratio, RPM, etc.) to minimize CO_2 emissions. The advantage of SVM is that accurate classification is possible even with fewer data points, and it can handle nonlinear patterns in combustion dynamics as well.

5.5 K-nearest neighbors

K-nearest neighbors (KNN) is a supervised learning algorithm mostly used for classification tasks. It is known as an instance-based learning algorithm, meaning it doesn't explicitly "learn" a model during training. Instead, it stores the entire dataset and makes predictions only when a new data point is given. Suppose we have a labeled dataset from solar panels categorized as "Faulty" or "Healthy" based on features such as panel temperature, voltage output, and current. When a new unlabeled data reading is received by the processor, the distance (difference) between the new point and all data points in the labeled dataset is calculated. Usually, Euclidean distance is used to measure the difference, but Manhattan distance or other metrics can also be applied. The closest data points in the dataset are then selected. Next, the labels of the selected data points are analyzed for majority voting to classify the panel accordingly. For example, if five data points are the closest and three of them are *Faulty*, then algorithm will classify the panel as "Faulty."

5.6 Extreme gradient boosting

Extreme gradient boosting (XGBoost) is based on the gradient boosting framework. It is widely used due to its speed, accuracy, and scalability. The boosting technique combines multiple weak learners to form a strong learner. The decision trees are built one after another, with each new decision tree minimizing the loss function using gradient descent. XGBoost is suitable for tabular data, such as a CSV file with structured data. We can use XGBoost to predict the output of multiple solar panels with varying efficiencies. Suppose a solar farm has multiple panels, but their efficiency varies due to shading (e.g., clouds, dust, or obstructions) and panel degradation over time. To increase efficiency, we aim to predict total energy output by intelligently combining weak signals from individual panels. Each solar panel provides weak predictive signals (e.g., voltage, current, temperature, irradiance). XGBoost combines multiple decision trees to correct errors from previous models. The first tree makes a rough prediction (e.g., "Panel 1 is underperforming due to shade"), and subsequent trees correct errors. The final model combines all weak signals into a strong, accurate forecast. For instance, if a panel is shaded, XGBoost may predict reduced output and adjust inverter settings to draw more energy from unshaded panels. Additionally, XGBoost may flag panels that are degrading faster than others.

5.7 Multilayer perceptron

A multilayer perceptron (MLP) is a type of feedforward neural network that consists of multiple layers of interconnected neurons. It is used for supervised learning tasks such as classification and regression. A neuron represents a feature or input variable of a neural network. The layers between the input and output are called hidden layers. As the number of layers increases, the prediction accuracy also increases, but it may also lead to overfitting. Hidden layers use nonlinear activation functions such as the Rectified Linear Unit, while the output layer uses a sigmoid function for binary classification and a SoftMax function for multiclassification.

5.8 K-means

K-means clustering is an unsupervised learning algorithm that groups a dataset into K distinct, nonoverlapping clusters based on feature similarity. The data points in the same cluster are more similar to each other than to those in other clusters. The similarity is measured by Euclidean distance, where the distance to a centroid data point is minimized. Let's go through a scenario involving an electricity grid that needs to balance supply and demand, especially with the rise of EV charging stations. K-means clustering helps identify patterns in EV charging behavior to optimize energy distribution. Sensors collect data on the time of charging (peak/off-peak hours), duration of charging, and energy consumption from the charging stations, and send it via an IoT network. K-means clustering groups similar charging profiles to predict high-demand hours.

5.9 Density-based spatial clustering of applications with noise

Density-based spatial clustering of applications with noise (DBSCAN) is an unsupervised algorithm used in ML and data mining. Unlike K-means, which is a centroid-based algorithm, DBSCAN groups data points based on their density distribution, making it effective for clusters of arbitrary shapes and for handling noise. Clusters are formed by core points and border points. A core point is a point that has a predefined minimum number of points within its neighborhood. A border point is a point that has a predefined threshold for border points, which is fewer than the number required for core points. All points that are neither core points nor border points are considered noise for that cluster. DBSCAN is not suitable for areas where the density of data points varies significantly.

5.10 Principal component analysis

Principal component analysis (PCA) reduces the number of features in a dataset while preserving the information. PCA standardizes the data for all the features, ensuring a mean of 0 and a variance of 1. Calculating the covariance matrix of the data reveals how features vary together, and its eigenvectors provide directions in space where the data vary the most or least. These directions are uncorrelated with each other. The corresponding eigenvalues indicate how much variation exists along each direction. The directions (eigenvectors) with the largest eigenvalues are selected to reduce the number of variables.

5.11 Autoencoders

The purpose of autoencoders is also to reduce the number of features but by using a different methodology. Autoencoders are used primarily for unsupervised learning tasks, such as dimensionality reduction and feature learning. They encode input data into a compressed representation and then decode it back to reconstruct the original input as closely as possible. For example, a high-resolution picture with millions of pixels can be reduced to a few thousand pixels by an autoencoder that still captures the essence of the image. Then, the decoder can reconstruct the original image from those few thousand values. The autoencoder is trained to reconstruct the image as close to the original image as possible. If the reconstruction is close to the original, it means the autoencoder has successfully learned the most important features.

5.12 Label propagation

One kind of ML technique that performs well when you have a lot of unlabeled data but little labeled data is label propagation. It operates on the straightforward premise that two data items most likely fall into the same category if they are quite similar. This is accomplished by treating each data point as a dot (node) in a network, with lines (edges) connecting the dots, whose strength indicates how similar the points are to one another. While the unlabeled dots begin with no label or have an equal probability of having every label that might be assigned, the dots that already have labels are kept. Labeled dots' label information is then regularly passed to their unlabeled neighbors by the algorithm, with stronger connections exerting greater effect. Based on the labels of the dots that are connected to it, each dot modifies its label. This process continues until there is little change in the labels. This method is particularly helpful for tasks such as categorizing documents into groups when the majority are unlabeled, identifying photos when labeling is costly, and classifying web pages with limited examples.

5.13 Self-training neural networks

A semisupervised learning technique called self-training neural networks allows a model to learn from its own predictions. There is a considerably larger set of unlabeled data at the beginning than there is labeled data. You start by using only the labeled data to train your neural network. After it has acquired knowledge, you utilize the trained network to predict labels for the unlabeled data. The model then treats these predictions, about which it is most certain, as though they were actual labels. The model is retrained using these newly "labeled" instances that have been added to the training set. The quantity of data that the model has "taught itself" from increases each time this process is repeated. Because it takes advantage of the vast amount of unlabeled data that is available, this approach is frequently employed in applications such as medical image classification, speech recognition, or text classification, where labeling data is costly or time-consuming.

5.14 Long short-term memory

Long short-term memory (LSTM) is a ML algorithm designed to overcome the limitations of the recurrent neural network (RNN). RNNs have a simple feedback loop that is sufficient for short sequences. However, when processing long sentences or time series, they suffer from the vanishing gradient problem. During training, gradients – that is, values used to update the network's weights – are calculated through backpropagation. In long sequences, these gradients can become too small, preventing the network from learning connections between distant events. As a result, the early parts of the sequence are lost during training. To solve this problem, LSTM is designed to allow information to remain effective over many time steps, enabling it to learn and remember long-range dependencies in sequential data.

6. Conclusion

The combination of ML with green energy and IoT technologies marks a significant advancement in creating more efficient, sustainable, and intelligent energy frameworks. Supervised learning is particularly effective for predictive modeling when labeled datasets are available, and unsupervised learning reveals hidden patterns in unlabeled information.

In renewable energy, especially within solar and wind energy systems, ML has proven to be vital for forecasting, oversight, and enhancement. IoT devices continuously gather data on temperature, irradiance, current, voltage, and environmental factors. ML algorithms analyze this information to identify faults, spot underperforming parts, and forecast future energy output. Techniques such as electroluminescence imaging, ultrasonic testing, and impedance spectroscopy are improved through ML models, decreasing dependence on manual inspections and facilitating predictive maintenance. This interconnection between IoT and ML enhances the efficiency of renewable energy systems while minimizing downtime and prolonging the lifespan of essential infrastructure.

The in-depth analysis of algorithms illustrates the wide-ranging applications of ML in green energy. Linear regression and logistic regression offer straightforward yet effective prediction and classification models. Decision trees provide rule-based decision-making, while random forests and gradient boosting improve predictive precision by mitigating overfitting and amalgamating weaker learners. Innovative techniques such as SVMs address nonlinear behaviors in energy systems, and KNN categorizes fault indicators in solar panels. Neural networks, such as MLPs and LSTMs, facilitate time-series prediction and long-term dependency modeling, making them crucial for forecasting demand and managing emissions. Unsupervised techniques such as k-means, DBSCAN, PCA, and autoencoders cater to anomaly detection, clustering of energy consumption patterns, and reducing the dimensions of high-volume sensor data. Semisupervised approaches such as label propagation and self-training neural networks capitalize on the wealth of unlabeled data, lessening the requirement for expensive manual labeling.

A notably significant use case is in emission management and combustion optimization. By employing ML algorithms to assess variables such as flame temperature, air-to-fuel ratio, and pressure, energy systems can be fine-tuned in real time to minimize CO_2 and NO_x emissions. Unlike traditional alert systems, which are inflexible and limited to a narrow range of features, ML-driven systems can simultaneously analyze numerous variables, adapt to fluctuating conditions, and generate more sophisticated predictions. This functionality positions ML not merely as a monitoring instrument but as a fundamental force for smarter, adaptive, and environmentally conscious energy management.

In summary, the examination of methodologies and applications clearly demonstrates that ML is indispensable for the future of renewable energy. As energy systems grow more intricate and data-intensive, conventional control methods struggle to cope with the sheer volume and variability of information. ML offers scalable solutions that enhance prediction accuracy, support predictive maintenance, optimize grid management, and reduce environmental effects. The integration of ML, IoT, and green energy heralds a new era of intelligent, robust, and sustainable energy systems, capable of meeting the increasing energy demands while effectively addressing climate change.

Author details

Waheeduddin Hyder[1*], Waqas Ahmed Siddique[2] and Mujtaba Sheikh[3]

1 Department of Computer Science, Millennium Institute of Technology and Entrepreneurship, Karachi, Pakistan

2 Department of Computer Science, DHA Suffa University, Karachi, Pakistan

3 Telecommunication Department, Quaid-e-Awam University of Engineering, Science and Technology, Nawabshah, Pakistan

*Address all correspondence to: fhyder152@gmail.com

References

[1] "CO2 Emissions – Global Energy Review 2025 – Analysis," IEA. [Online]. Available: https://www.iea.org/reports/global-energy-review-2025/co2-emissions Accessed: July 14, 2025

[2] "Annual change in energy-related CO2 emissions, 1900-2024 – Charts – Data & Statistics - IEA." [Online]. Available: https://www.iea.org/data-and-statistics/charts/annual-change-in-energy-related-co2-emissions-1900-2024 [Accessed: 2025-July-14]

[3] Różycki R, Solarska DA, Waligóra G. Energy-aware machine learning models—A review of recent techniques and perspectives. Energies. 2025;**18**(11):2810. DOI: 10.3390/en18112810

[4] Bolón-Canedo V, Morán-Fernández L, Cancela B, Alonso-Betanzos A. A review of green artificial intelligence: Towards a more sustainable future. Neurocomputing. 2024;**599**:128096. DOI: 10.1016/j.neucom.2024.128096

[5] Hua J, et al. Artificial intelligence for calculating and predicting building carbon emissions: A review. Environmental Chemistry Letters. 2025;**23**(3):783–816. DOI: 10.1007/s10311-024-01799-z

[6] Aslam S, Aung PP, Rafsanjani AS, Majeed APPA. Machine learning applications in energy systems: Current trends, challenges, and research directions. Energy Informatics. 2025;**8**(1):62. DOI: 10.1186/s42162-025-00524-6

[7] Hajri NHA, Harthi RNA, Pasam GK, Natarajan R. IoT and machine learning based green energy generation using hybrid renewable energy sources of solar, wind and hydrogen fuel cells. E3S Web Conference. 2024;**472**:01008. DOI: 10.1051/e3sconf/202447201008

[8] Emexidis C, Gkonis P. The integration of Internet of Things and Machine learning for energy prediction of Wind Turbines. Applied Sciences. 2024;**14**(22):10276. DOI: 10.3390/app142210276

[9] Singh R, Akram SV, Gehlot A, Buddhi D, Priyadarshi N, Twala B. Energy system 4.0: Digitalization of the energy sector with inclination towards sustainability. Sensors. 2022;**22**(17):6619. DOI: 10.3390/s22176619

[10] Rahman MS, Ghosh T, Aurna NF, Kaiser MS, Anannya M, Hosen ASMS. Machine learning and internet of things in Industry 4.0: A review. Measure: Sensors. 2023;**28**:100822. DOI: 10.1016/j.measen.2023.100822

[11] "Application of Machine Learning to predict CO2 Emissions in Light-Duty Vehicles." [Online]. Available: https://www.mdpi.com/1424-8220/24/24/8219 [Accessed: 2025-September-04]

[12] Zhou K, Fu C, Yang S. Big data driven smart energy management: From big data to big insights. Renewable and Sustainable Energy Reviews. 2016;**56**:215–225. DOI: 10.1016/j.rser.2015.11.050

[13] Directive 2010/75/EU of the European Parliament and of the Council of 24 November 2010 on industrial emissions (integrated pollution prevention and control) (recast) (Text with EEA relevance), vol. **334**.

2010. [Online]. Available: http://data.europa.eu/eli/dir/2010/75/oj/eng [Accessed: 2025-July-18]

[14] Roibas-Millan E, Cubas J, Pindado S. Studies on cup anemometer performances carried out at IDR/UPM Institute. past and present research. Energies. 2017;**10**(11):11. DOI: 10.3390/en10111860

[15] "Figure 2 Sonic Anemometer/MTi-G schematic. Note that U axis of the. . .," [Online]. ResearchGate. Available: https://www.researchgate.net/figure/Sonic-Anemometer-MTi-G-schematic-Note-that-U-axis-of-the-anemometer-is-aligned-with-the_fig2_258748292 [Accessed: 2025-August-24]

[16] Olabi AG, Wilberforce T, Elsaid K, Salameh T, Sayed ET, Husain KS, Abdelkareem MA. Selection Guidelines for Wind Energy Technologies. Energies. 2025;**14**:3244. DOI: 10.3390/en14113244

[17] Shelby Z, Hartke K, Bormann C. The Constrained Application Protocol (CoAP). Internet Engineering Task Force, Request for Comments RFC 7252. 2014. DOI: 10.17487/RFC7252

[18] Duffie JA, Beckman WA. Solar Engineering of Thermal Processes. New Jersey: John Wiley & Sons; 2013. DOI: 10.1002/9781118671603

Chapter 4

Energy Storage Systems: Bridging the Gap Between Supply and Growing Demand

Anna Pinnarelli, Pasquale Vizza and Alireza Soleimani

Abstract

The rapid growth in global energy demand, driven by urbanization, industrial expansion, and electrification, poses critical challenges to the stability, flexibility, and sustainability of modern energy systems. Simultaneously, the increasing integration of non-dispatchable renewable energy sources introduces significant variability and intermittency, threatening grid reliability. Energy storage systems (ESS) have emerged as a key enabler in addressing these issues by decoupling generation and consumption, enhancing grid stability, and facilitating higher penetration of renewable energy. This chapter provides a comprehensive overview of ESS technologies, including electrochemical, mechanical, thermal, and hydrogen-based systems, alongside their technical characteristics, maturity levels, carbon footprint, and economic considerations. It compares short-term versus long-term storage solutions and evaluates their strategic roles in supporting daily operations, seasonal balancing, and sector coupling. The chapter also highlights global trends in energy consumption, the challenges of renewable integration, and the need for hybrid storage architectures. Ultimately, it demonstrates that ESS is not merely a technical solution but a systemic pillar for enabling a resilient, low-carbon, and equitable energy transition.

Keywords: energy storage systems, grid flexibility and stability, decarbonization, short-term and long-term storage, sustainable energy transition

1. Introduction

Energy systems are facing major challenges due to the ongoing rise in global energy demand and the increasing integration of renewable energy sources. Energy storage systems (ESS), which improve grid flexibility, dependability, and efficiency, are essential in tackling these issues. ESS aids in balancing supply and demand, reducing grid congestion, and decarbonizing the energy sector by storing excess energy and delivering it during periods of high demand. Emerging technologies, including advanced batteries, thermal storage, and hydrogen-based solutions, offer scalable and sustainable options to meet growing energy needs. Despite facing technical, economic, and regulatory barriers, innovations in storage technology and supportive policies are accelerating ESS deployment, making it a key enabler for a sustainable and resilient energy future.

1.1 The global energy demand

Global energy demand trends highlight a structural tension between growing consumption and environmental sustainability goals. According to the most recent analyses by the International Energy Agency (IEA), overall energy demand is expected to continue rising in the coming years, driven primarily by emerging economies, where urbanization, industrialization, and rising levels of prosperity are significantly increasing energy needs. At the same time, high-income countries are seeing a relative stabilization of demand, driven by energy efficiency, technological innovation, and the progressive electrification of consumption. This multi-speed dynamic raises crucial questions: How can we ensure a secure and affordable supply while simultaneously reducing dependence on fossil fuels? The energy transition, in this sense, is not just a technological transformation but a systemic process that requires coherence between public policies, industrial investments, and social justice mechanisms aimed at ensuring that the most vulnerable communities are not excluded.

In 2024, global energy demand increased by 2.2%, which was more than the average growth rate over the previous ten years. In 2024, demand increased for all fuels and technologies. Due to record temperatures, electrification, and digitization, the demand for energy increased by 4.3%, which was significantly more than the 3.2% growth in global GDP. The power sector was the main driver of this increase. With 38% of the expansion in the world's energy supply coming from renewable sources, natural gas came in second with 28%, coal with 15%, oil with 11%, and nuclear with 8% [1].

1.2 Energy transition and the role of storage

The increasing penetration of non-programmable RES and the phase-out of part of the thermoelectric capacity will lead to a further deterioration of both the adequacy and safety conditions of the system, as well as frequency stability. Currently, power systems are also exposed to increasingly stressful situations due to severe climatic conditions (extreme heat or cold, poor water supply) and/or, as happens in those countries from which Italy imports energy, the eventual unavailability of nuclear power plants. The impact that non-programmable RES can have on system safety can be amplified in weakly interconnected systems or small islanded systems, which are characterized by moderate energy requirements and a production capacity concentrated in a few large and inflexible production plants. The principal impacts of all the previously described aspects on the system are as follows: the reduction of power reserve margins, the increase in congestion on the electricity grid (due to the uneven distribution of RES throughout the territory), the lessening of the system inertia and damping, the greater need for resources capable of providing regulation services (frequency and voltage) in a short time, and, in general terms, a greater need to move resources on the Dispatching Services Market. By projecting the criticalities that are already evident today into the near future, we are facing an epochal challenge. Accounting for over-generation too (in which renewable production will exceed consumption), this will be fully evident when the penetration of production from non-programmable RES covers all the needs, at least in the central hours of the day.

Consequently, to achieve the decarbonization goals defined for 2030 and 2050, specific intervention plans must be developed and implemented at the system level. According to [2], global energy storage capacity will increase by 35% in 2025, reaching 94 GW (247 GWh), excluding pumped hydro. This marks the highest annual increase ever recorded [2].

The same source forecasts a compound annual growth of 14.7% through 2035, with annual additions reaching 220 GW (972 GWh) by that year.

By 2030, the worldwide energy storage capacity is expected to reach 1,500 GW, six times the 2022 level, according to the COP29 Energy Storage and Grids Pledge.

Power supplies and demands must be balanced by storage devices. Rapid energy exchange between storage systems and the grid may be possible. This trend is consistent with machine mechanics' intervention times for the damping and inertia properties (synthetic inertia/damping contribution). The system providing grid services (voltage and frequency regulation) needs storage capacity that can help maintain frequency stability. Additionally, it ought to help ensure that energy generated by utilizing non-programmable RES can be stored.

ESSs and the local dispersed RES can work together to generate synergies. An additional function of resources placed for other purposes, the synthetic inertia/damping service, is based on energy storage devices. As a result, many services may not be able to be performed concurrently by the same storage system. A service may request energy creation at one moment and energy absorption at another. Different services could need varying degrees of the storage system's charge state to function at any one time. The predicted benefits of many services operating simultaneously must, therefore, be carefully considered when weighing the costs and benefits of storage systems.

Determining the ideal mix and, thus, the ideal storage capacity set aside for the provision of synthetic inertia and damping is crucial. Rapid, power-intensive reaction during high RES generation and energy-intensive response to feeding local electrical loads, even in the absence of RES production, are the two main functions that ESSs must perform. However, it is challenging for a single type of storage device to fulfill these two functions. By combining the unique features of several storage device types, hybrid ESSs may offer an answer.

Storage systems are generally chosen based on the function they are designed to perform. For applications requiring large amounts of power exchanged with the electrical grid for short periods of time, for example, it will be necessary to use storage technologies that operate in power, while technologies that operate in energy will be chosen where the storage of large amounts of energy is required for longer periods of time.

- Very short-term storage applications (power quality, energy recovery system, UPS). These include storage systems capable of intervening in a very short time, even if with limited power levels, to support the voltage quality of privileged users. This category also includes storage systems capable of injecting or withdrawing large amounts of energy for very limited periods (e.g., braking energy recovery in electric vehicles). This category includes high-power supercapacitors and SMES, technologies characterized by low levels of total accumulable energy and still high costs per kWh.

- Applications supporting grid transients, for load/production shifting and the consequent increase in self-consumption. This storage category is the fastest growing, including those storage systems that primarily perform load-shifting functions (e.g., a storage system combined with a renewable energy plant allows for self-consumption of the energy produced and stored during hours of source availability, during hours when the source is unavailable but the load is present). This category also includes energy-intensive applications, such as the installation of ZEBRA batteries in distribution and transmission networks to smooth out peaks in production (typically from wind farms) or load demand following a lack of production.

- Grid support applications. This category includes high-power installations supporting grid voltage and frequency regulation. They, therefore, play a role similar to that offered by traditional thermoelectric plants. Currently, pumped storage basins are used exclusively for this purpose, with CAES systems being used less frequently. Clearly, the significant impact of this type of storage system on the landscape and territorial context slows the construction of new installations and, consequently, the expansion of the market. In the medium term, this category could grow with the addition of mixed virtual units, energy communities, and, more specifically, Community Energy Storage. Thanks to new regulations regarding "virtual aggregation," the aggregation of centrally managed Community ESS will be able to function as distributed storage, offering both grid support services – overcoming the landscape and territorial constraints of large-scale plants – and the functions described in the previous category.

1.3 Objectives and scope of the chapter

The primary objective of this chapter is to provide a comprehensive analysis of the role of energy storage in addressing the challenges posed by the increasing global demand for energy. The chapter explores the different technologies available – ranging from electrochemical batteries and pumped hydro storage (PHS) to thermal and hydrogen-based systems – highlighting their respective advantages, limitations, and fields of application. Furthermore, it discusses their role in smart grid management and decarbonization, highlighting what can influence the deployment of ESS at both centralized and distributed scales in several case studies and real-world applications. Innovations and emerging trends in ESS are reshaping the energy landscape by enhancing efficiency, reducing costs, and enabling deeper integration of renewable resources into modern power grids.

The scope of the chapter extends beyond the technical characterization of storage systems, encompassing their integration into smart grids, their contribution to resilience in critical infrastructures, and their potential role in enabling sector coupling between electricity, heating, and mobility. By doing so, the chapter seeks to illustrate how energy storage is not only a technical solution but also a systemic driver for sustainable development, supporting both climate mitigation objectives and equitable access to reliable energy.

2. The growing challenge of energy consumption

Due in major part to population increase, industrial expansion, and improved living standards globally, energy consumption has been rising continuously for decades. Rapid industrialization and urbanization, particularly in emerging economies, greatly increases their energy requirements. Existing energy infrastructures, which frequently find it difficult to keep up with rising consumption while preserving affordability and dependability, are put to the test by this spike in demand. At the same time, technological advancements like the electrification of transportation and the digitization of services are making patterns of energy usage more complex.

Meeting growing demands while minimizing the environmental impact of energy production and use is a twofold problem brought on by increased energy demand. Fossil fuels continue to be the world's primary energy source, and they play a significant role in climate change and greenhouse gas emissions. Moving away from carbon-intensive energy sources becomes crucial as the world commits to aggressive climate targets, including those outlined in the Paris Agreement. However, especially in less developed areas, this shift needs to be handled cautiously to prevent jeopardizing economic growth and energy security.

The increasing electrification of many industries, transportation, and heating is another significant factor that raises the demand for energy. Electrification makes it possible to integrate renewable energy sources, which helps achieve decarbonization targets, but it also poses new challenges for grid stability and management. The varying output of renewable energy sources and fluctuating patterns of energy consumption necessitate flexible and adaptable energy systems that can instantly balance supply and demand.

One essential way to deal with these issues is to use ESSs. ESSs work to reduce fluctuations and improve grid resilience by storing excess energy during times of low demand and releasing it during times of peak consumption. Furthermore, storage technologies promote grid stability and reduce the intermittency of renewable energy sources, allowing for a greater penetration of these sources. Building a sustainable and resilient energy future requires a comprehensive strategy that integrates energy efficiency, renewable deployment, and sophisticated storage solutions in order to successfully address the growing energy demand dilemma.

2.1 Trends in global energy use

Global energy consumption has increased during the last 20 years due to changes in geopolitical environments, technical advancements, and socioeconomic progress. The IEA reports that between 2000 and 2024, total final energy consumption increased by more than 25%, with non-OECD nations accounting for the majority of this rise. Industrialization, growing middle classes, and initiatives to extend energy access are all contributing to the sharp rise in the demand for electricity in emerging economies in Asia, Latin America, and Africa. In contrast, energy consumption in high-income nations has either decreased or plateaued, mostly as a result of increased electrification of end-use sectors, structural changes in the economy, and advancements in energy efficiency. The decoupling of energy demand from economic growth in a number of industrialized economies is one of the most important phenomena of the past several years. Because of higher building standards, more efficient appliances, and the digital revolution of sectors, energy consumption has either remained relatively flat or decreased even while GDP continues to expand. In many emerging regions, where infrastructure is still being created and energy intensity is still high, this decoupling is not yet apparent. Energy demand is predicted to increase further as these regions pursue economic development, especially in metropolitan areas, which are forecast to house almost 70% of the world's population by 2050. As transportation, heating, and industrial processes continue to be electrified, electricity is emerging as the primary energy carrier. As electric cars, heat pumps, data centers, and smart technologies proliferate, electricity demand is predicted to continue growing. In 2024, it increased by 4.3%, surpassing the rise of the world GDP. Although this change offers opportunities to include sustainable energy sources, it also necessitates major improvements to demand-side management (DSM), energy storage capacity, and grid infrastructure in order to maintain stability and effectiveness. Furthermore, the energy mix is gradually changing, with a greater proportion of supply coming from renewable sources. In 2024, 38% of the increase in the world's energy supply came from renewable sources, with coal, oil, and natural gas coming in second and third, respectively. Although wind and solar PV are spearheading this change, the supply and demand balance is made more difficult by their unpredictability. These patterns highlight how urgently flexible, robust, and decarbonized energy systems are needed, and technologies such as demand response (DR), digital grids, and energy storage will be crucial in determining how the world's energy consumption develops in the future.

2.2 Urbanization, electrification, and industry growth

One of the main causes of the global growth in energy demand is urbanization. Urban centers are growing in size and population at a rapid rate as more people move to cities in pursuit of greater economic opportunities and higher living standards. As a result of this expansion, there is an increased need for energy-intensive public services, infrastructure, housing, and transportation. Nearly 70% of people on Earth are expected to live in cities by 2050, up from about 57% at the present time. To ensure efficiency, dependability, and sustainability in the face of this urban transformation, updating urban energy systems is just as important as increasing the energy supply.

As part of the global energy transition and decarbonization targets, important sectors such as transportation, heating, and industrial processes are becoming more electrified. Fossil fuel-powered alternatives are gradually being replaced by electric furnaces, electric buses, electric heat pumps, and electric cars. This change greatly raises electricity usage, even though it lowers local emissions and makes it easier to integrate renewable energy. For instance, by 2050, the electrification of transportation alone is predicted to increase the world's electricity demand by up to 4,000 TWh per year, which is equal to the EU's current total electricity consumption. In order to effectively handle additional loads, this trend emphasizes the significance of energy storage, smart charging infrastructure, and grid upgrading.

Global energy consumption continues to be significantly influenced by industrial growth, especially in emerging economies. Energy-intensive industries, including manufacturing, steel, cement, and chemicals, are growing to keep up with demand worldwide. Many developing regions are still in the early stages of industrialization, when energy demand is naturally higher per unit of GDP, while some industrialized nations are moving toward service-based economies. By using cleaner fuels, electrifying industrial processes, and integrating cutting-edge technologies such as waste heat recovery, carbon capture, and hydrogen, the aim is to decouple industrial growth from emissions.

Urbanization, electrification, and industrial growth all put pressure on energy production, distribution, and consumption, necessitating a strategic rethink. Deploying adaptable, scalable, and low-carbon energy solutions is necessary in light of these dynamics. ESS can be crucial in this regard since they facilitate peak shaving, enable grid flexibility, and guarantee the consistent supply of energy in both urban and industrial environments. ESS technology will be essential for load balancing, integrating dispersed renewable sources, and creating resilient energy infrastructure in the future as cities and businesses grow more energy-intensive.

2.3 Renewable integration and intermittency issues

In order to decarbonize the energy system, the global energy transition is increasingly depending on the widespread use of renewable energy sources, such as hydropower, wind, and solar photovoltaics. By providing sustainable, clean substitutes for fossil fuels, these technologies greatly lessen their negative effects on the environment and greenhouse gas emissions. However, integrating them into current energy systems presents a number of operational and technological difficulties, the most significant of which is the intermittency problem. The inherent variability and weather dependence of solar and wind generation, in contrast to conventional power plants, make it more difficult to ensure a steady and uninterrupted energy supply.

Grid reliability is seriously threatened by intermittency, particularly in systems that rely heavily on renewable energy sources. For example, solar electricity is only available during the day and may vary depending on cloud cover, while wind power is influenced by the unpredictable patterns of the wind. If not well controlled, these variations may result in frequent supply and demand mismatches, which could lead to frequency aberrations, voltage instability, and even blackouts. In order to stabilize the system in real time, grid operators are now depending more and more on flexible resources, including DR, energy storage, and fast-ramping power plants.

The phenomenon of overgeneration at times of strong RES output and low demand, especially during midday hours in solar-rich countries, is a significant obstacle to renewable integration. In certain situations, it might be necessary to limit renewable energy in order to preserve grid balance, which would result in energy waste and decreased economic efficiency. In contrast, extra capacity from dispatchable sources or stored energy must be promptly available to satisfy demand during times of low RES output, such as at night or during calm, overcast weather. In order to preserve stability and maximize the penetration of renewable energy sources, this emphasizes the necessity of grid flexibility and sufficient backup systems.

By separating the timing of energy generation and consumption, ESSs are essential for reducing the unpredictability of renewable energy sources. ESSs can eliminate the requirement for fossil-based peaker plants, smooth out output variability, and increase grid stability by storing excess energy during periods of peak production and releasing it during periods of demand or renewable deficits. Furthermore, enhanced storage options like pumped hydro, battery storage, and newer technologies like green hydrogen improve grids' capacity to handle larger proportions of renewable energy sources while lowering curtailment and bolstering ancillary services. Therefore, resolving intermittency is a strategic facilitator for the complete integration of renewables in the global energy environment, rather than just a technical necessity.

3. Fundamentals of ESS

3.1 Definition and classification of ESS

An ESS is a technology that enables the temporal decoupling of energy production and consumption by storing excess energy at a given moment and releasing it later when demand requires it or when production is insufficient. This seemingly simple concept plays a crucial role in ensuring continuity, stability, and sustainability in modern energy systems.

ESSs operate at the intersection of three fundamental dimensions:

1. Technological: related to the type of conversion and storage adopted.

2. Economic: concerning investment costs, operating expenses, and the lifetime.

3. Systemic: referring to the role that storage plays within the broader energy ecosystem.

ESSs can be classified at several levels:

- By the form of stored energy:

  - Electrochemical: rechargeable batteries (lithium–ion, sodium–sulfur, flow batteries). They offer high efficiency and modularity but are constrained by critical raw materials and degradation phenomena.

  - Mechanical: solutions that store energy as potential energy (pumped hydro, compressed air energy storage (CAES)) or kinetic energy (flywheels). Suitable for large-scale, long-duration storage but often limited by geographical constraints.

  - Thermal: based on sensible heat (water tanks), latent heat (phase change materials – PCM), or thermochemical storage. Particularly useful for sector coupling between electricity and heating.

  - Direct electrical: capacitors, supercapacitors, and superconducting magnetic energy storage (SMES). Extremely fast but with limited capacity.

- By power and discharge duration:
  - High power – short duration: designed for instantaneous grid regulation and transient compensation.
  - High energy – long duration: designed to cover energy deficits over days or even seasons.
- By functional application:
  - Ancillary services → frequency control, voltage regulation.
  - Energy shifting → transferring energy from low-demand to high-demand periods.
  - Backup and resilience → ensuring service continuity in case of outages.
  - Sector integration → bridging electricity, heat, and mobility.

The value of storage lies not only in its technical capacity but also in the flexibility it introduces into the system. Depending on the scale, ESSs can provide:

- Distributed storage: residential and commercial systems reducing grid dependence.
- Centralized storage: large-scale installations supporting national power systems.
- Hybrid architectures: combining different technologies to maximize performance and minimize costs.

3.2 Comparison of short-term vs. long-term storage

The comparative analysis between short-term and long-term storage is fundamental to understanding how ESSs meet diverse yet complementary needs.

Short-term storage solutions primarily serve dynamic grid management. Key features include the following:

- Speed and accuracy: capable of reacting to load or generation fluctuations within milliseconds or seconds.
- High cycling capability: frequent charge and discharge cycles (hundreds or thousands per year) with manageable performance degradation.
- Modular flexibility: easily deployable in decentralized contexts, such as homes, commercial buildings, or small-scale PV plants.
- Limitations: the main drawback lies in the relatively small storage capacity and higher cost per kWh, which make them less suitable for prolonged energy supply.

From a system perspective, short-term storage is essential for the following:

- Ensuring power quality and stability,
- Reducing the risk of local blackouts,
- Enhancing self-consumption in microgrids.

Long-term storage, by contrast, serves a strategic role in ensuring continuity over medium to long time frames.

- Seasonal balancing: In renewable-dominated systems, solar and wind production often diverge from demand curves. Long-term storage enables surplus energy accumulation during periods of abundance and release during shortages.
- Large-scale deployment: typically implemented at a regional or national scale, with capacities on the order of GWh or even TWh.
- Efficiency and costs: Although less efficient than short-term solutions, their systemic value lies in replacing fossil-fuel backup plants.
- Technological constraints: Many options require dedicated infrastructures (dams, salt caverns, electrolysis, and methanation plants) and substantial capital investment, making large-scale deployment both a technical and economic challenge.

The contrast between short-term and long-term storage can be further assessed along three main parameters:

1. Cost per unit of stored energy (€ per kWh):
   a. Lower for long-term storage at a large scale, though it requires a high upfront investment.
   b. Higher for short-term solutions, but with modular and scalable deployment.
2. Lifetime (years or cycles):
   a. Batteries are limited by their finite cycle life and gradual degradation.
   b. Mechanical systems (PHS, CAES) typically last several decades with routine maintenance.
3. Strategic role within the energy system:
   a. Short-term → daily stability management, peak shaving, ancillary services.
   b. Long-term → seasonal balancing, security of supply, and pathway to full decarbonization.

Synthetic comparison		
Characteristic	**Characteristic**	**Characteristic**
Time horizon	Seconds–hours	Hours–months
Round-trip efficiency	80–95%	30–80%
Typical capacity	kWh–MWh	GWh–TWh
Response time	Milliseconds–minutes	Minutes–hours
Main technologies	Batteries, supercapacitors, flywheels	Pumped hydro, CAES, hydrogen
Typical applications	Frequency regulation, ancillary services, daily peak shaving	Seasonal balancing, strategic reserve

Table 1.
Comparative overview of multi-scale energy storage technologies.

Future energy systems will not rely on a single technology. Instead, they will require hybrid architectures that combine:

- Fast, high-efficiency solutions for daily and intraday balancing.
- Large-scale, long-duration systems for seasonal and strategic storage.

This multi-scale energy storage paradigm is considered key to building a resilient and sustainable infrastructure, capable of harmonizing intermittent renewable generation, increasingly electrified end-use sectors, and a smart, digitalized power grid (**Table 1**).

3.3 Technology readiness level (TRL) and maturity of ESS

The deployment potential of energy storage technologies is strongly linked to their TRL, which measures the degree of technological maturity on a scale from laboratory concept (TRL 1–3) to full commercial deployment (TRL 9).

- High TRL – commercially mature technologies:
    - PHS: TRL 9, the most established storage solution worldwide, represents more than 90% of installed capacity.
    - Lithium–ion batteries: TRL 9, widely adopted in both stationary storage and electric mobility, with rapidly decreasing costs and continuous improvements in energy density and safety.
    - Lead-acid batteries: TRL 9, mature but gradually declining due to lower efficiency and shorter lifespan compared to lithium-based systems.
- Medium TRL – emerging but already demonstrated technologies:
    - Flow batteries (vanadium redox, zinc-bromine): TRL 6–8, promising for long-duration storage thanks to decoupled energy and power scaling, but still limited by high costs and material availability.

- Sodium–sulfur (NaS) batteries: TRL 7–8, demonstrated in several utility-scale projects, with challenges related to operating temperature and safety.
- CAES: TRL 7–8, operational in some large-scale plants, with further development required for adiabatic and hybrid configurations.

- Low TRL – early-stage or pilot technologies:
  - SMES: TRL 4–6, excellent response time, but limited scalability and high cryogenic costs.
  - Thermochemical storage: TRL 4–6, significant potential for seasonal applications, but it is still at the research or demonstration stage.
  - Hydrogen and power-to-X solutions: TRL 5–7, promising for long-term and cross-sector storage, but requiring cost reductions in electrolysis, fuel cells, and infrastructure.

In summary, PHS and lithium–ion batteries dominate the current market, while flow batteries, CAES, and hydrogen represent the frontier of long-duration storage, with ongoing R&D efforts to achieve cost competitiveness and operational reliability.

3.4 Comparison of energy storage technologies based on carbon footprint and O&M costs

Beyond technical performance, two crucial factors in evaluating ESS are the carbon footprint and the operation and maintenance (O&M) costs over the system's lifetime. These parameters directly influence the sustainability and economic viability of storage deployment (**Table 2**).

Carbon footprint

- Lithium–ion batteries: Manufacturing involves the mining and processing of critical minerals (lithium, cobalt, nickel), which contribute significantly to

Comparative table			
Technology	**Carbon footprint (life cycle)**	**O&M costs**	**Lifetime**
Lithium–ion battery	Medium (critical materials)	Medium	8–12 years
Lead-acid battery	High	High	3–7 years
Flow battery	Medium–low	Low	15–20 years
PHS	Low (construction phase)	Very low	40–60 years
CAES	Medium (fuel-dependent)	Medium	25–40 years
Hydrogen/power-to-X	Variable (depends on source)	High	20–30 years

Table 2.
Operational and environmental profiles of key ESS.

embodied emissions. The life cycle carbon footprint is mitigated by high efficiency and potential recycling, but concerns remain about supply chains and material scarcity.

- Lead-acid batteries: higher emissions per unit of stored energy due to lower efficiency and shorter lifetimes. Widely recyclable, yet the environmental burden of lead processing is significant.
- Flow batteries: moderate carbon footprint, with emissions concentrated in electrolyte production and stack components. Their longer lifetime and ability for deep cycling reduce environmental impact on a per-kWh basis.
- Pumped hydro storage (PHS): Very low operational emissions and a long lifetime (>40 years). The carbon footprint mainly arises from civil works (dams, reservoirs) during construction.
- CAES: Carbon intensity depends on configuration. Conventional CAES, using natural gas, emits significantly, while advanced adiabatic CAES can achieve near-zero operational emissions.
- Hydrogen (Power-to-gas, power-to-X): High carbon footprint if hydrogen is produced from fossil fuels (gray hydrogen). When coupled with renewable electrolysis (green hydrogen), life cycle emissions decrease drastically, though current efficiency levels remain low.

O&M Costs

- Lithium–ion batteries: Moderate O&M costs, primarily related to battery management systems (BMS) and cell degradation monitoring. Replacement cycles are required after 8–12 years, depending on usage.
- Lead-acid batteries: Low upfront cost, but high O&M costs due to frequent replacements and reduced depth of discharge.
- Flow batteries: Higher initial capital cost, but O&M is relatively low due to their long cycle life and modular electrolyte replacement.
- PHS: Low O&M costs relative to capacity, with maintenance mainly focused on turbines, pumps, and civil infrastructure. A long service life translates to a very favorable levelized cost of storage (LCOS).
- CAES: Moderate O&M costs, linked to turbomachinery maintenance. Fuel costs remain relevant in conventional CAES systems.
- Hydrogen-based storage: Currently, high O&M costs are due to inefficiencies in electrolysis, compression, and reconversion. These costs are expected to decrease with technological learning curves and large-scale deployment.

Table 2 compares the operational and environmental profiles of ESS.

4. Role of ESS in smart grid management and decarbonization

Following the examination of the electrical storage technologies described so far, in which the applicability and convenience of each technology for applications related to the energy sector and energy communities have been highlighted, it is considered useful to provide an overview of the electrical storage technologies in relation to the categories of applications. Particularly interesting for this purpose is the mapping provided by IRENA [3] in **Figure 1**.

4.1 Grid stabilization and load balancing

Energy storage systems stabilize and balance the grid by absorbing excess electricity and releasing it when needed, providing rapid response to frequency fluctuations and supporting renewable energy integration. They also manage peak demand by storing off-peak energy for peak hours, reduce grid strain by alleviating congestion, and can offer grid services such as voltage support and black start capabilities, enhancing overall grid reliability and resilience.

4.1.1 How storage systems stabilize the grid?

Frequency regulation: ESS can quickly inject or absorb power to maintain stable grid frequency, which is crucial for preventing outages and improving resilience, especially with intermittent renewable sources.

Voltage support: They provide active control over voltage, contributing to a stable and consistent power flow across the grid.

Ancillary services: ESS offer various ancillary services, including spinning reserve (a form of backup power) and black start capability, which help restore critical grid infrastructure after a blackout.

4.1.2 How storage systems balance the grid?

Demand response and peak shaving: ESS store energy during low-demand (off-peak) hours and discharges it during high-demand (peak) hours, flattening the overall demand curve and reducing the need for expensive peak power plants.

Renewable energy integration: They store surplus energy from renewable sources such as wind and solar when production is high and release it when generation is low, smoothing out their inherent variability and enabling greater integration.

Grid decongestion: By storing and distributing energy locally, ESS reduce strain on transmission and distribution networks, helping to defer costly infrastructure upgrades.

Energy arbitrage: ESS can store energy when it is cheap and sell it back to the grid when prices are high, making the grid more efficient and cost-effective.

4.2 Peak shaving and DR

Storage systems are a versatile technology that provide both peak shaving and DR capabilities, helping to reduce energy costs, improve grid stability, and integrate renewable energy sources more effectively.

Peak shaving is a calculated tactic that enables users to maximize their energy use by reducing their electricity consumption during times of high demand. These times are usually marked by an increase in energy consumption, which raises prices and

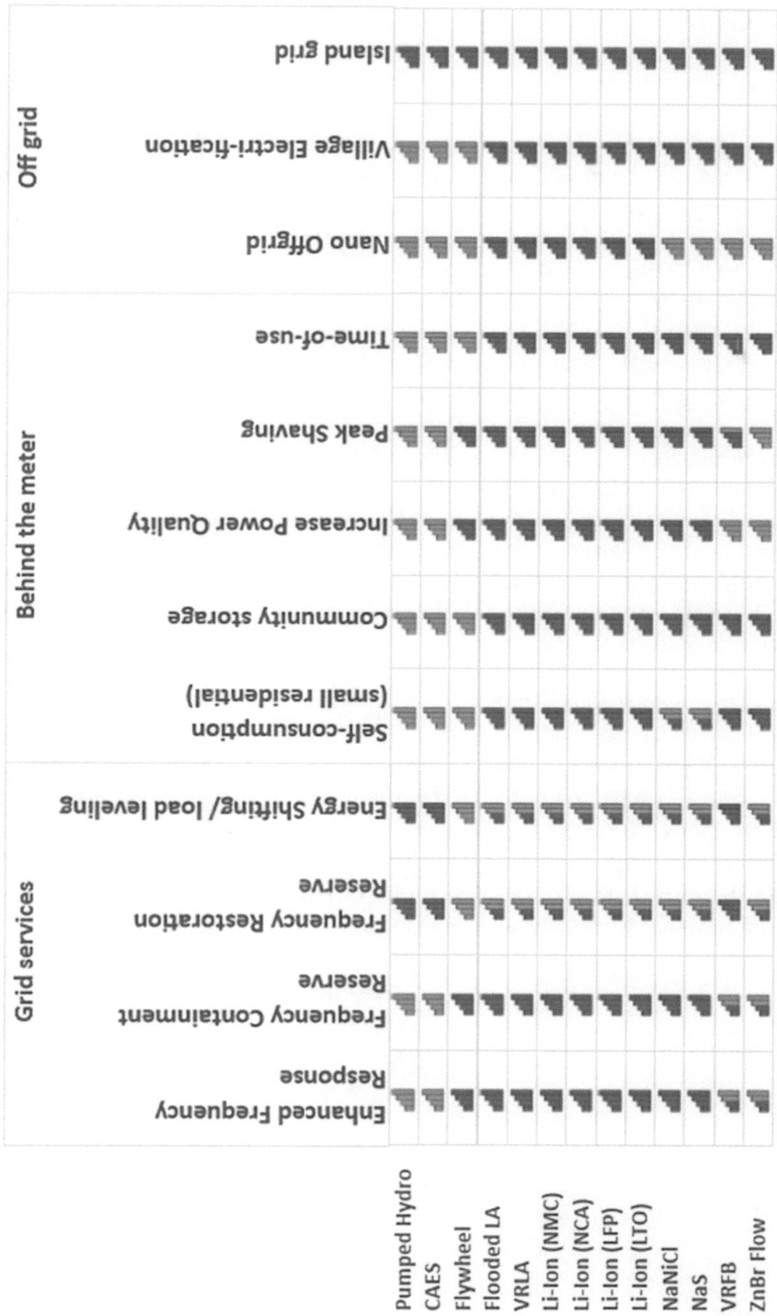

Figure 1.
Technology mapping of electrical storage systems for application categories [3].

may put stress on the electrical grid. In this process, energy storage devices are essential because they serve as a buffer by holding extra electricity during off-peak hours and releasing it during peak hours, which lessens dependency on the grid. By stabilizing the grid, supplying energy when needed, and facilitating the incorporation of renewables, they also help with DR.

4.2.1 How storage systems enable peak shaving?

These systems charge up by absorbing excess energy from the grid or renewable energy sources like solar panels during times of low energy demand, when electricity prices are generally lower. The stored energy is seamlessly released during periods of high demand, offering a dependable and affordable substitute for drawing power directly from the grid. In addition to lessening the burden on the grid, this process enables customers to benefit from more affordable electricity rates and avoid high-demand fees.

By adopting storage systems for peak shaving, businesses can lower electricity bills, avoid costly demand charges, and contribute to grid stability by balancing the overall electricity grid. Additionally, batteries improve energy reliability and provide backup power during outages, which is critical for high-energy industries like manufacturing plants or data centers (**Figure 2.**).

Examples: In an EV charging site with limited grid supply, a battery can provide supplementary power during periods of high charging demand and then recharge when demand is low. **Figure 2.** explains peak shaving.

ESS can help businesses optimize their energy usage and participate in DR programs. These systems not only help businesses lower costs but also contribute to grid stability and sustainability goals.

4.2.2 How storage systems enable DR?

Load shifting: A technique known as storage system load shifting involves an ESS, like a BESS, charging with grid electricity during off-peak, cheaper times and then discharging that stored energy during peak, more expensive times. By shifting the timing of energy use, this optimizes energy use to reduce end-user power bills, ease the strain on the grid during periods of high demand, and allow for a deeper integration of renewable energy sources (**Figure 3.**).

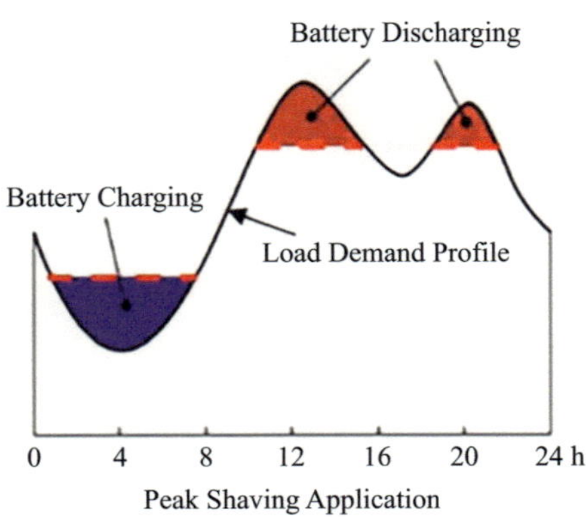

Figure 2.
Enabling peak shaving and DR through ESS [4].

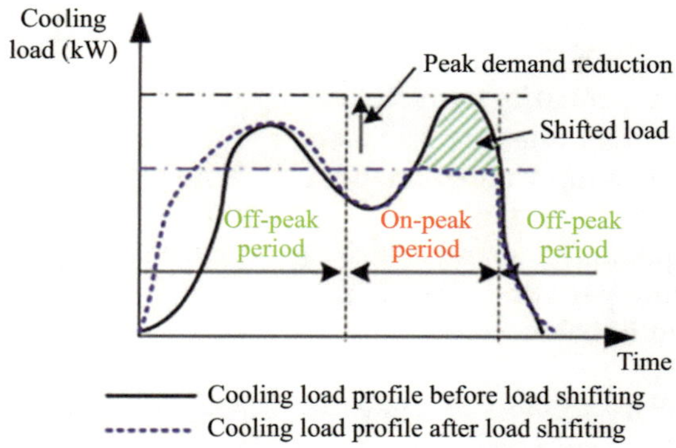

Figure 3.
Enhancing DR capabilities through load shifting with energy storage [5].

The key differences between the two approaches are as follows:

Peak shaving: Focuses on reducing energy consumption from the grid during specific high-demand peak periods by using stored energy.

DR: A broader strategy that involves changing energy consumption patterns over time, often by shifting usage to off-peak hours. While peak shaving is a component of DR, DR can also involve other techniques, such as load shedding or shifting.

4.3 Integration with renewable energy sources

The intermittency of renewable energy sources, such as solar and wind, is mitigated by storage system integration, which provides grid stability and reliability by storing excess energy during peak output and releasing it during low periods [6, 7].

Smart grids, DSM strategies, and sophisticated control systems are typically included in a comprehensive integration approach. Given the latest developments in these technologies, we will touch on these points in passing in this section [8].

The adoption of a hybrid energy storage system (HESS) is required since the various ESS technologies currently on the market are unable to meet the wide frequency spectrum of the generated energy. Two complementary storage devices that can be connected in a variety of topologies typically make up a HESS [9, 10].

4.4 Enhancing grid stability, reliability, and resilience

Improving the stability, reliability, and resilience of the power grid is essential to accommodate increasing renewable energy integration and evolving demand patterns. Grid stability ensures consistent voltage and frequency levels, preventing disturbances that can cause outages or equipment damage. Reliability focuses on the continuous availability of electricity, minimizing interruptions for consumers and critical infrastructure. Resilience emphasizes the grid's ability to withstand, recover, and adapt to extreme events such as storms, cyberattacks, or equipment failures. Implementing advanced monitoring systems, smart grid technologies, and energy storage solutions plays a crucial role in strengthening these aspects, allowing the grid to respond dynamically to fluctuations in supply and demand while maintaining secure and uninterrupted service. explains flexibility in power systems.

4.4.1 Grid stability

Storage solutions, primarily in the form of batteries, pumped hydro, and thermal storage, can dramatically improve grid stability by addressing the challenges posed by fluctuating energy supply and demand. Here's how:

- Balancing supply and demand → Storage solutions can store excess energy generated during periods of low demand and release it when demand peaks, preventing grid imbalances.

- Providing ancillary services → They can provide essential services such as frequency regulation and voltage support, which are vital for maintaining grid stability. Frequency regulation ensures the grid operates at the correct frequency (typically 50 or 60 Hz), while voltage support helps maintain stable voltage levels.

- Integrating renewable energy → Storage solutions help smooth out the intermittent nature of renewable energy sources like solar and wind, making them more reliable and predictable.

Storage solutions balance electricity supply and demand, improving grid stability by providing ancillary services and integrating renewable energy sources.

Improving grid stability through storage solutions translates into numerous benefits:

- Reduced blackout risk → A stable grid is less prone to blackouts, ensuring a reliable power supply for homes, businesses, and critical infrastructure.

- Lower energy costs → Storage can help reduce peak demand charges and optimize energy use, leading to lower electricity bills.

- Increased renewable energy integration → A stable grid allows for greater integration of renewable energy sources, reducing reliance on fossil fuels and lowering carbon emissions.

- Enhanced grid resilience → Storage solutions enhance the grid's ability to withstand disruptions caused by extreme weather events, cyberattacks, or other unforeseen circumstances.

In essence, storage solutions act as a buffer, mitigating the inherent variability of both energy supply and demand, ultimately strengthening the grid's resilience and promoting a more sustainable energy future.

One of the most significant ways energy storage enhances reliability is by providing essential grid services. These services help maintain the stability and integrity of the power grid, ensuring that electricity flows smoothly and efficiently.

Some critical grid services include the following:

- Frequency regulation: The grid operates at a specific frequency (e.g., 50 Hz in Europe). Energy storage can quickly respond to fluctuations in frequency, injecting or absorbing power as needed to keep the grid stable.

- Voltage support: Maintaining stable voltage levels is crucial for preventing equipment damage and ensuring the proper operation of electrical devices. Energy storage can provide reactive power to support voltage levels, particularly in areas with weak grid infrastructure.

- Black start capability: In the event of a widespread blackout, energy storage can be used to restart the grid, providing the initial power needed to bring generators back online.

4.4.2 Grid reliability

Energy storage improves grid reliability by:

- Providing backup power during outages.

- Smoothing out fluctuations caused by renewable energy sources.

- Deferring or avoiding costly upgrades to transmission and distribution infrastructure.

- Offering grid services, such as frequency regulation and voltage support.

Energy storage functions as a power reservoir, capturing excess energy for deployment during peak demand, ensuring a stable and secure energy supply.

4.4.3 Grid resilience

Energy storage acts as a safety net, stepping in when traditional power sources falter and bolstering the grid's ability to handle unexpected events.

- Frequency regulation → Storage systems can rapidly respond to changes in grid frequency, maintaining stability.

- Black start capability → Some storage systems can restart parts of the grid after a total blackout, accelerating recovery.

- Peak shaving → Storage can reduce demand during peak hours, alleviating stress on the grid and lowering electricity costs.

Energy storage in microgrids ensures that critical facilities maintain power even when the main grid fails, a vital component of community resilience.

5. Innovations and emerging trends

Innovation in energy storage technology is accelerating at a never-before-seen rate as the energy sector undergoes a significant upheaval propelled by decentralization, digitization, and decarbonization targets. Next-generation ESSs are pushing the limits of what is theoretically and financially possible, thanks to developments in materials science, sophisticated control algorithms, and system integration. In addition to enhancing performance indicators such as energy density, longevity, and response time, these developments are opening up new storage applications at the utility, commercial, industrial, and residential levels.

The development of battery technologies beyond conventional lithium–ion chemistries is one of the most noticeable phenomena. Alternative chemistries such as sodium–ion, solid-state batteries, and flow batteries are becoming more popular, but lithium–ion still dominates the market because of its high efficiency, falling costs, and maturity. For example, sodium–ion batteries have cost and supply chain benefits, particularly in areas where cobalt and lithium are scarce. By substituting solid materials for liquid electrolytes, solid-state batteries offer increased energy density and enhanced safety. Redox flow batteries are appealing for long-duration storage applications because of their remarkable scalability and cycle life, despite their lower energy density.

Modular and hybrid storage systems are also becoming a major area of innovation. Fast frequency response and long-duration energy shifting are two benefits that HESS can provide by fusing several storage technologies, such as high-power lithium–ion systems with high-energy flow batteries. Because of their adaptability, they are perfect for intricate grid services where a single technology might not be enough. Particularly in decentralized and off-grid applications, modular storage units with plug-and-play architectures are facilitating quicker deployment and simpler integration into current infrastructures.

Simultaneously, ESS management and operation are being transformed by digitization and AI. Real-time optimization of charging and discharging cycles by intelligent algorithms has increased system longevity and efficiency. Machine learning makes predictive maintenance possible, which lowers operating costs and improves dependability. Additionally, with the help of blockchain technology and sophisticated energy management systems, the integration of storage into smart grids, peer-to-peer energy markets, and virtual power plants (VPPs) is growing quickly. These systems enable dispersed storage assets to engage in dynamic interactions with other grid components, offer ancillary services, and participate in energy markets.

In the future, multi-vector energy systems and sector coupling will be at the forefront of energy storage innovation. Holistic decarbonization is being facilitated by the integration of energy, heating, transportation, and industrial processes, made possible by adaptable storage technologies such as green hydrogen and thermal storage. In particular, hydrogen has a lot of potential as a clean energy carrier and long-duration storage medium that can help offset seasonal imbalances between demand and renewable supply. Globally, governments and industry players are investing more in these cutting-edge technologies, indicating a clear move toward energy systems that are more robust, adaptable, and sustainable.

5.1 AI and IoT in storage management

The management, optimization, and integration of ESSs into contemporary energy networks are being completely transformed by the combination of AI and the Internet of Things (IoT). Traditional management techniques are no longer adequate as energy systems become more complex due to decentralized generation, variable renewable energy sources, and changing demand patterns. The transition to intelligent, data-driven energy storage operations is made possible by AI and IoT technologies, which improve system responsiveness, efficiency, and reliability.

IoT technologies enable data collection and real-time monitoring throughout the storage system. Storage systems can gather a great deal of information about battery state-of-charge, state-of-health, temperature, voltage, current, and usage cycles via embedded sensors, edge devices, and smart meters. AI systems process and evaluate this fine-grained data to produce actionable insights on centralized or cloud-based platforms. A more open and flexible energy storage ecosystem that can react instantly to customer and grid demands is the end outcome.

Furthermore, the development and management of VPPs, which combine and manage dispersed storage assets typically found in residences, workplaces, and industrial facilities, are made possible by AI-IoT integration. These cloud-coordinated, intelligent systems can offer useful grid services such as peak shaving, voltage control, and frequency regulation. Additionally, AI-powered storage systems allow dynamic load balancing when paired with renewable forecasting and DR, which lessens the need for backup production based on fossil fuels.

5.2 Second-life batteries and recycling

A significant number of lithium–ion batteries are nearing the end of their initial life due to the rapid expansion of electric cars and renewable energy sources. Many of these batteries still have a substantial capacity and can be utilized again in second-life applications, such as stationary energy storage, even though they are no longer suitable for high-performance applications. In addition to storing renewable energy and supporting grid stability, these systems can often save costs by as much as 40% compared to new batteries. Although second-life batteries offer a sustainable option, there are still issues with performance inconsistency, lack of standardization, and safety concerns. Large-scale deployment requires addressing these problems. Recycling is essential for recovering important materials such as lithium, cobalt, and nickel once batteries can no longer be reused. The development of technologies like hydrometallurgical and direct recycling is increasing the efficiency and environmental friendliness of material recovery. Efficient recycling promotes supply chain resilience and reduces reliance on mining.

5.3 Hybrid energy storage systems (HESS)

In order to benefit from their complementary properties, HESs incorporate two or more distinct energy storage technologies. By merging several storage systems into a single system, this technique aims to overcome the disadvantages of each one and improve overall performance, dependability, and efficiency. Depending on the needs of the application, common combinations include batteries with flywheels, CAES, or supercapacitors.

In energy systems, hybrid storage is vital for minimizing the unpredictability and intermittent nature of renewable energy sources such as solar and wind. For example, supercapacitors offer high power density and rapid charge/discharge capabilities, while batteries offer high energy density, ideal for long-term energy storage. These technologies can be coupled to build hybrid systems that provide a consistent supply of energy as well as a quick dynamic reaction to variations in load or grid disturbances.

The management and control of hybrid storage systems involve complex energy management strategies that optimize the power flow between different storage units. This ensures maximum efficiency, prolongs component lifespan, and maintains system stability. Advanced control algorithms continuously monitor parameters such as state of charge, power demand, and system health to determine the optimal allocation of energy between storage devices. Hybrid storage systems have found increasing applications in microgrids, electric vehicles, and large-scale grid support. Their ability to balance short-term power demands with long-term energy requirements makes them indispensable in modern energy infrastructure, particularly as the penetration of renewable energy sources continues to rise. Consequently, hybrid storage is considered a key enabler for sustainable and resilient energy systems.

6. Conclusion(s)

Energy storage systems have emerged as a cornerstone technology in the evolving landscape of global energy systems. As energy demand continues to rise, driven by urbanization, industrial growth, and electrification, ESS play a vital role in ensuring system flexibility, stability, and resilience. Their ability to decouple generation from consumption allows for efficient integration of intermittent renewable energy sources, mitigates grid instability, and supports the transition toward decarbonized energy systems. While technical, economic, and regulatory barriers still exist, ongoing innovation and supportive policy frameworks are rapidly advancing storage deployment. To achieve a secure, sustainable, and low-carbon energy future, a diversified mix of short-term and long-term storage solutions tailored to specific applications and integrated within smart energy systems will be essential.

Conflict of interest

The authors declare that there is no conflict of interest regarding the publication of this chapter.

Acknowledgment

The authors acknowledge the use of Grammarly for language polishing of the manuscript.

DOI: http://dx.doi.org/10.5772/intechopen.1013978

Author details

Anna Pinnarelli*, Pasquale Vizza and Alireza Soleimani
Department of Mechanical Energy and Management Engineering (DIMEG), University of Calabria, Rende, Italy

*Address all correspondence to: anna.pinnarelli@unical.it

References

[1] International Energy Agency (IEA). Global Energy Review 2025 [Internet]. Paris: IEA; 2025. Available from: https://www.iea.org/reports/global-energy-review-2025. Licence: CC BY 4.0 [Accessed: 2025-October-30]

[2] Nsitem N, Sekine Y. Global energy storage growth upheld by new markets. BloombergNEF. 2025 Jun 18. Available from: https://about.bnef.com/insights/clean-energy/global-energy-storage-growth-upheld-by-new-markets/ [Accessed: 2025-October-30]

[3] International Renewable Energy Agency (IRENA). Electricity Storage and Renewables: Costs and Markets to 2030. Abu Dhabi: IRENA; 2017

[4] Ananda-Rao K, Ali R, Tanisellas S. Battery energy storage system assessment in a designed battery controller for load levelling and peak shaving applications, Journal of Renewable and Sustainable Energy, 2017;**9**(4). DOI: 10.1063/1.4991455

[5] Sun Y, Wang S, Xiao F, Gao D. Peak load shifting control using different cold thermal energy storage facilities in commercial buildings: A review. Energy Conversion and Management. 2013;**71**:101–114. DOI: 10.1016/j.enconman.2013.03.026

[6] Ould Amrouche S, Rekioua D, Rekioua T, Bacha S. Overview of energy storage in renewable energy systems. International Journal of Hydrogen Energy. 2016;**41**(45):20914–20927. DOI: 10.1016/j.ijhydene.2016.06.243

[7] Soleimani A, Roghanian P, Heidari M, Heidari M, Pinnarelli A, Vizza P, Mahdavi M, Mousavi S.F. Refining hybrid energy systems: Elevating PV sustainability, cutting emissions, and maximising battery cost efficiency in remote areas. Energy Nexus. 2025;**18**:100452

[8] Malik FH, Hussain GA, Alsmadi YMS, Haider ZM, Mansoor W, Lehtonen M. Integrating energy storage technologies with renewable energy sources: A pathway toward sustainable power grids. Sustainability. 2025;**17**:4097. DOI: 10.3390/su17094097

[9] Etxeberria A, Vechiu I, Camblong H, Vinassa JM. Hybrid energy storage systems for renewable energy sources integration in microgrids: A review. In: Proceedings of the 2010 Conference on Power and Energy (IPEC 2010); Singapore. New York: IEEE; 2010. p. 532–537. DOI: 10.1109/IPECON.2010.5697053. [Accessed: 2010-October-27–29]

[10] Jeon S, Bae S. Integrated optimisation for sizing, placement, and energy management of hybrid energy storage systems in renewable power systems. Journal of Energy Storage. 2025;**106**:114793

Chapter 5

Alternative Propulsive and Drivetrain Methodologies for Energy Engineering in Agricultural and Off-Road Vehicles

Elijah Collins, Reece Durbin, Nicholas Hoff, Brandon Kitch, Porter Little, Tovia Owens, Seshan Calapatti Suresh and Robert M. Stwalley III

Abstract

This chapter is a review of the potential alternative propulsion power and drivetrain methods in agricultural and off-road vehicles. It is vital that sustainable, efficient, and resilient solutions to off-road vehicle power sources be found to overcome the economic, environmental, and operational challenges associated with the transition from fossil fuel sources. The areas to be explored in this chapter are electric vehicles, diesel–electric hybrids, hydrogen fuel cell and internal combustion engines, natural gas and methane engines, and biofuel-powered tractors, as well as electric-continuously variable transmissions and hydrostatic transmissions. Each topic will be highlighted along with research and analysis pertaining to the feasibility, advantages, and disadvantages of each topic, in comparison to their more traditional alternatives. The findings of each alternative propulsion and traction method will be further evaluated to determine recommendations moving forward pertaining to the manufacture of each method, or if the method should be reconsidered for use in modern-day agriculture and off-road vehicles. Recommendations will divide the examined technologies into those with immediate potential, those that may be useful in the near future, and those unlikely to be of use within the agricultural realm for some time to come.

Keywords: all-electric, diesel–electric hybrid, hydrogen engine, hydrogen fuel cell vehicle, methane engine, biofuels, electric-continuously variable transmission, hydrostatic transmission

1. Introduction

The internal combustion engine (ICE) with #2 diesel oil as its fuel has been the dominant power source in agriculture since the 1960s. As reducing greenhouse gas emissions and pollution becomes an ever more prominent concern, it is important to

consider possible alternatives for powering the equipment of the agricultural industry. Ongoing technological progress has created potential new and alternative sources of power for off-road vehicles, which may provide energy sources different than fossil fuels. However, the adoption of these alternative methodologies faces significant barriers: high upfront costs, limited infrastructure, variable energy storage performance, and the need for durability under rugged field conditions. Additionally, research on alternative propulsion and drivetrain technologies for agricultural and off-road applications remains fragmented, with insights scattered across mechanical, electrical, and agricultural engineering domains.
A comprehensive synthesis of these emerging technologies is needed to clarify their technical feasibility, operational trade-offs, and long-term potential. Industry stakeholders, policymakers, and researchers need a clearer understanding of the pathways available to transition toward more sustainable, resilient, and efficient power systems for agricultural and off-road vehicles. This paper provides a review of the implications of these novel technologies powering agricultural equipment, focusing primarily on all-electric vehicles, diesel–electric hybrids, hydrogen fuel cell and ICEs, methane engines, and biofuel-powered tractors. Along with describing possible alternative power sources, this chapter also aims to explore some different types of transmissions, such as E-CVTs and hydrostatic transmissions (HSTs), comparing them to conventional mechanical transmissions. The practical, economical, technological, and feasible advantages and limitations of each technology will be compared to the conventional diesel engine or conventional method of drivetrain power transfer. Literary reviews and analyses for each power source and transmission type will be provided along with a concluding overview.

2. Electric propulsion variants

All-electric tractors represent a unique innovation in sustainable agriculture, combining environmental stewardship with advanced technology. The concept dates back to the late nineteenth century, with early experiments, such as the Zimmermann electric ploughs in 1894 and the Brutschke electric tractors in 1900. These early all-electric models faced challenges, including cumbersome power supplies and limited battery technology, which hindered widespread adoption [1]. However, recent advances in battery technology and an increased focus on reducing greenhouse gas emissions have revitalized interest in electric tractors. These modern electrical machines offer several advantages over their diesel counterparts, including lower maintenance costs due to fewer moving parts, reduced vehicular emissions, and fuel cost savings [2, 3].

Several companies are leading the way in the development of all-electric tractors. Solectrac®, based in California, launched its first electric tractor model in 2017, designed for small farms and vineyards [4]. Monarch Tractor®, also from California, introduced a fully electric, driver-optional smart tractor aimed at enhancing both sustainability and efficiency [5]. Meanwhile, established agricultural manufacturers such as New Holland have developed advanced models, such as the T4 Electric Power® utility tractor, which features autonomous capabilities and zero vehicular emissions [6]. The adoption of electric tractors is further supported by government incentives and grants aimed at promoting sustainable farming practices. These initiatives help farmers offset the higher initial costs associated with transitioning to electric machinery, making it more accessible for a wider range of agricultural operations [7]. The evolution of all-electric tractors reflects the agricultural sector's shift toward more sustainable and efficient practices. By overcoming historical challenges and leveraging modern advancements, electric tractors are poised to play a pivotal role in shaping the future of farming.

All-electric tractors are increasingly being recognized as a viable alternative to traditional diesel-powered models, offering significant advantages in terms of performance, cost efficiency, and environmental sustainability, particularly for utility and animal feedlot mission profiles. Recent studies indicate that all-electric tractors also perform competitively in field operations, while charged and operational. For instance, an electric all-wheel-drive tractor demonstrated robust performance during plowing, maintaining an average voltage of 68.2 *V* and a motor torque of 79.4 *Nm*, with a maximum traction power of 42.0 *kW* [8]. Similarly, a 25 *hp* distributed drive electric tractor (DDET) achieved a traction force of approximately 5,400 *N* and accelerated from 0 to 28.2 *km/h* in just 5.5 *s*, highlighting its efficiency for dynamic agricultural tasks [9]. Operational longevity, or range, remains a significant issue for all-electric tractor field operations.

From a cost perspective, electric tractors can offer significant savings over their life cycle. The primary operational cost, electricity, can be as low as $12 to fully charge a 100 *kWh* battery, assuming an electricity rate of $0.12/*kWh* [7]. Moreover, a comparative analysis of a compact electric tractor (CET) and a 32 *hp* diesel tractor found that, over time, the CET offers cost advantages due to lower energy and maintenance expenses, despite the higher initial purchase price [10]. These factors contribute to a lower total cost of ownership compared to traditional diesel models, making electric tractors an economically attractive option for long-term use, when the machines are in an appropriate mission profile that does not require extended propulsive field work activities.

The environmental benefits of all-electric tractors are also noteworthy. By aligning with global sustainability goals and producing zero tailpipe emissions, these machines may reduce greenhouse gas output, if electricity generation paradigms can be changed to renewable sources. Additionally, all-electric tractors with fewer moving parts require less maintenance than their diesel counterparts, which reduces downtime and further enhances their operational efficiency [9]. However, the transition to electric tractors is not without challenges. The need for adequate charging infrastructure and adjustments in farm operations to accommodate charging times are critical considerations for widespread adoption. Despite these challenges, the advantages of electric tractors in reducing emissions, lowering operational costs, and improving farm efficiency position them as a transformative solution in the agricultural industry.

The operational performance of all-electric tractors is largely determined by the capabilities of their battery systems and motor technologies. Recent advancements have enhanced their tractive efficiency, allowing all-electric tractors to deliver sufficient torque and traction power for most agricultural applications. For example, DDETs demonstrate robust dynamic performance, achieving acceleration exceeding traditional diesel models, while providing high torque output at low speeds. This makes them suitable for tasks such as plowing and hauling [9]. Electric tractors also excel in precision farming, as their consistent torque and quiet operation allow for better integration with advanced automation and sensor systems.

Electric tractors significantly reduce environmental impact, producing zero on-site emissions and helping to improve local air quality. Unlike hydrogen, methane, or biofuel systems, they operate without combustion, eliminating on-site CO_2 and NO_x emissions. These benefits make electric tractors especially appealing in regions with stringent emissions regulations. However, their environmental advantage depends on the energy mix used for charging. In regions where electricity generation relies heavily on fossil fuels, the net environmental benefits may be diminished or even reversed [8].

Cost is another critical consideration. Although electric tractors have higher initial purchase prices compared to diesel-powered models, their operating costs are substantially lower. Electricity is generally cheaper than diesel fuel on a kWh basis, and the reduced number of moving parts minimizes maintenance requirements. Over time, these savings can offset the initial investment, especially for farms with access to renewable energy sources for charging. However, the limited battery range, often between 4 and 10 hours per charge, poses challenges for extended or high-demand operations, requiring careful planning to avoid downtime during critical farming periods [10].

Infrastructure requirements remain a significant barrier to widespread adoption. Many rural areas lack the charging infrastructure needed to support large-scale electric tractor usage. Additionally, the relatively long charging times compared to refueling diesel vehicles may disrupt time-sensitive agricultural tasks. Advances in fast-charging technology and the integration of renewable energy sources on farms could mitigate some of these challenges in the future [7].

Overall, electric tractors show promise as a sustainable alternative to traditional propulsion methods, particularly in regions with established charging infrastructure and renewable energy integration. However, their feasibility depends on continued advancements in battery technology, cost reduction, and infrastructure development to address operational limitations. Additionally, the issue of rapid battery depletion during heavy operations poses a significant roadblock to all-electric tractor use for field activities.

The diesel–electric hybrid is a greener alternative to traditional diesel engines, but it does not have the limiting issues of the all-electric tractor, which will slow its wide-scale adoption. The diesel–electric hybrid has been recently introduced into the agricultural and off-road machinery market, but it is not yet widely used. It is projected that this alternative power method will quickly gain popularity [11]. The advantages of the diesel–electric hybrid engine are substantial compared to others on the market or in the laboratory. One benefit of this design is the efficiency that it has over gasoline–electric hybrid engines, since the diesel engine has an improved fuel economy over a gasoline engine. A diesel–electric design also reduces fuel consumption levels, and therefore, it reduces emissions compared with traditional diesel engines. It provides a reduction in fuel waste when at idle engine speeds and an increased runtime, when compared to a fully electric vehicle that requires downtime for charging [12]. Torque outputs are also higher and quicker to respond to operator input. The disadvantages that come from the diesel–electric hybrid engine are a higher cost and an increased overall machine weight, due to the additional electric components required [13]. According to the European Inland Barging Innovation Platform (2024), for water-based transportation, "a hybrid system reduced fuel consumption, in some occasions as much as 25%, and therefore, it increased the range of the vessel." [11].

There have been both historic and current research around the idea of diesel–electric hybrid engines, and the implementation of this concept is on-going and has governmental support [14]. For example, locomotive engines equipped with internal combustion diesel engines are almost universally electrically driven [15]. Although this is different from a diesel–electric hybrid engine, it still exemplifies the way in which mechanical power generated by a diesel engine can be turned into electrical power through an alternator, which can then power electric motors to drive the locomotive's axles. In an actual hybrid, the power generated by the diesel engine can be stored in batteries and then converted into electric power via electric motors [16]. A more current example of research in the area of diesel–electric hybrids is the AbramsX tank developed by General Dynamics in collaboration with the US Army [17]. This new tank design offers many benefits that the traditional tanks in the army are not equipped with, including lighter weight, less required operator input, and increased fuel efficiency [18].

Using a diesel–electric hybrid design, the diesel engine operates within its optimal speed range, powering the generator to produce electricity. This power can either directly drive the electric motors or be stored in the batteries for later use, allowing for flexible power distribution based on demand. Modern diesel–electric hybrids often incorporate regenerative braking systems, which can recover up to 25–30% of the energy typically lost during braking [19]. This recovered energy is stored in the batteries and can be used later for acceleration or auxiliary power needs [20]. The outlook for diesel–electric hybrid engines is very positive in the future, as these engines offer multiple advantages from efficiency and emissions standpoints, but they are also very robust, power dense, and not reliant directly upon the power grid to charge their batteries. Bunton et al. (2025) explored this option and its considerable benefits for agricultural tractors in detail [21].

3. Hydrogen propulsion

Hydrogen vehicles in existence today operate on one of two principles: fuel cells or internal combustion. In a fuel cell electric vehicle (FCEV), hydrogen is introduced to oxygen in a special reaction chamber and the resultant chemical reaction generates electricity, which is then used to power the vehicle. They are effectively electric vehicles directly using hydrogen fuel and oxygen from the air as their power source [22]. The Apollo spacecraft was provided with electrical power by hydrogen fuel cells [23]. Hydrogen can also be used as a supplement or sole fuel source for the ICE heat release process. This configuration is noted for having a lower efficiency than FCEV power sources at light or medium loads, but at higher or full loads, which are likely for agricultural, construction, and mining purposes, these engines perform better than FCEVs [22]. There are a number of considerations when incorporating hydrogen into an internal-combustion engine, and this has been an area of on-going research since at least 1977 [24]. Fuel storage is also a significant concern, as hydrogen is very diffusive and difficult to contain. The high autoignition temperature and low ignition energy are design considerations which must be addressed early in specific engine system development. The high autoignition temperature can introduce issues for diesel engine-based designs, leading to it often being used in conjunction with diesel fuel for pilot ignition or traditional spark plugs [25].

There are a number of benefits to hydrogen-powered vehicles, namely their clean burning nature, high efficiency, and the abundance and renewable nature of the basic fuel source. The by-product of hydrogen-powered vehicles is water vapor, and water is the most abundant molecule on the surface of the planet by mass. However, there are significant drawbacks to hydrogen power as the technology exists today. Given that it is fairly new, there are limited infrastructure facilities available to support it, meaning that there are few places to refuel. Another concern which is frequently raised in discussions of hydrogen power is safety. It is highly flammable, as the *Hindenburg* proved [26]. Safety issues are currently not fully resolved, but the problems are not intractable.

There are some hydrogen vehicles currently available or under development for both automotive and agricultural purposes. New Holland has partnered with Blue Fuel Solutions to create their H2 Dual Power® tractor, while First Mode is a green energy company that is currently providing solutions to modify existing heavy machinery for mining to improve efficiency and increase electrification. New Holland's tractor runs with dual fuel. It can be run with a combination of hydrogen and diesel fuel, or diesel alone [27]. One of First Mode's options is a hydrogen FCEV system for agricultural tractors [28]. Additionally, Deutz has constructed a six-cylinder hydrogen engine that it plans to begin mass production of in 2025 [29]. In addition to those current engines now in production, a number of companies and organizations are looking into producing hydrogen-powered off-road equipment. Cummins is working on a proof-of-concept hydrogen-burning ICE [30]. CAT is planning a hydrogen-hybrid ICE [31], and CNH has developed multiple early-phase demonstrators and continues to pursue hydrogen FCEV tractors [32].

Available studies on hydrogen-fueled ICE performance frequently utilize four-stroke, single-cylinder diesel engines. There are many variations in these studies, observing the effects of varying amounts of hydrogen gas, different methods of fuel injection, and performance differences when hydrogen is mixed with various alternative fuels. Gomes Antunes et al. (2009) contrasted hydrogen direct injection (DI) with standard diesel performance, homogenous charge compression ignition, and hydrogen blended with diesel as a pilot fuel [33]. They used an air-cooled, naturally aspirated engine with a 17:1 compression ratio. They also implemented heating of the inlet air to account for the increase in ignition delay shown by hydrogen fuels at low temperatures. This allowed for stable engine operation in the DI mode, something Gomes Antunes et al. (2008) had difficulty with using a homogeneous charge compression ignition engine [34]. The DI engine also had a 20% reduction in NO_x compared with the diesel. The results obtained for each of the four engines are shown in **Table 1**.

	Diesel	Diesel + Hydrogen	Homogeneous charge compression injection	Direct injection
Shaft output %	27.9	33.9	48.0	42.8
Cooling system %	42.2	31.2	20.4	17.3
Exhaust gases %	35.3	34.9	31.6	39.9
Shaft power (*kW*)	9.0	8.9	7.1	10.3

Table 1.
Comparison of results from testing various hydrogen utilization schemes in engines [33, 34].

Hydrogen %	0.0	16.0	36.0	46.0
Indicated thermal efficiency %	43.48	41.98	40.67	38.91
Specific fuel consumption (*g/kWh*)		201.8	208.3	217.7
CO_2 emission (*g/kWh*)	622.5	520.9	351.5	235.4
CO emission (*g/kWh*)	1.70	0.56[a]	0.56[a]	0.56[a]
THC (*g/kWh*)	0.017	0.031	0.028	0.030
Smoke (*g/kWh*)	0.66	0.28	0.20	0.16
NO_x (% increase)	0	0	58.8	159.7
Peak cylinder pressure (*bar*)	53.4	57.5	66.5	72.9
Peak heat release rate (*J/°C*)	30.3	35.7	45.9	52.0

[a]*Lower limit of measurement.*

Table 2.
Comparison of engine metrics from various hydrogen fuel levels [35].

Sandalci and Karagöz (2014) varied the rate of hydrogen ingestion in a dual-fuel diesel engine [35]. They used a water-cooled, naturally aspirated, single-cylinder, four-stroke engine with a variable compression ratio. The compression ratio was set to 19:1 to mimic diesel performance, while engine torque and speed were maintained at specific settings. Cylinder pressure, thermal efficiency, specific fuel consumption, power, and emissions were recorded. **Table 2** shows a compilation of their results.

Bose et al. (2013) used a single-cylinder, four-stroke, water-cooled, direct-injection diesel engine and compared its performance under varying loads and durations of hydrogen injection, using a diesel–hydrogen blend with the original diesel engine's performance characteristics [36]. They observed an increase in brake thermal efficiency compared to diesel fuel in the hydrogen–diesel blend under all loads, with a peak value of 30% improvement at maximum load. They also observed reduced brake specific fuel consumption (BSFC) and brake specific energy consumption values at all loads, and they postulated that this was due to improved combustion from the hydrogen. Senthil Kumar et al. (2003) evaluated the performance of a single-cylinder, direct-injection, four cylinder, water-cooled diesel engine with a mixture of vegetable oil and hydrogen [37]. They compared it with the use of vegetable oil alone as a fuel, as well as hydrogen with diesel as the pilot fuel [37]. Using both the vegetable oil and the diesel fuel, they found that introducing increasing amounts of hydrogen raised and then lowered brake thermal efficiency at increasingly high loads. Hydrogen use reduced smoke in Bosch smoke units (BSU), hydrocarbon (HC), and CO emissions. It increased NO_x emissions, ignition delay, and it decreased combustion duration. At optimized hydrogen inclusion, 7% by mass, brake thermal efficiency increased from 27.3% to 29.4% using vegetable oil. At 5% hydrogen by mass, brake thermal efficiency increased from 30.3% to 32% with diesel fuel. **Table 3** presents these results from varied fuel mixes.

The performance of hydrogen ICEs showed significant variation and was highly dependent on the fuel blend and injection method. Hydrogen-supplemented ICEs can be designed to have higher brake thermal efficiencies, high shaft power output, and acceptable levels of ignition delay. Hydrogen-supplemented engines generally had significantly reduced CO_2, CO, and smoke emissions, and they burned more

	% brake thermal efficiency	Smoke (BSU)	HC (*ppm*)	CO (%)	NO (*ppm*)	Ignition delay (°*C*)
Vegetable oil	27.3	4.4	130	0.26	735	10
Vegetable oil + 7% hydrogen	29.4	3.7	100	0.17	875	13
Diesel fuel	30.3	3.9	100	0.20	785	9
Diesel fuel + 5% hydrogen	32.0	2.2	70	0.10	894	11

Table 3.
Comparison of engine metrics at various hydrogen fuel levels using vegetable oil and diesel fuel as the main fuels in a diesel engine [37].

completely. However, most hydrogen engines do have some level of engine knock and tend to have increased NO_x emissions. Some initiatives are currently working to address the NO_x issue [38], while there are others focusing on improving the ignition delay in hydrogen-supplemented engines [39]. Hydrogen can ultimately be sourced through renewable or traditional means, but current sourcing techniques eliminate the potential environmental benefits of its use. These issues will need to be addressed before hydrogen-based options provide an environmentally sustainable fueling alternative for agricultural vehicles. In the various analyses, hydrogen was noted as having several roadblocks to become widely feasible, the most significant of which were the need for a stronger distribution infrastructure and better storage solutions. Hydrogen-based ICE units seem more immediately feasible than FCEVs, since the fuel cell technology reaction chambers are not yet optimized as a mass production item.

4. Natural gas and methane propulsion

Methane utilization is a growing alternative fueling source for the modern diesel-style engine. Methane, in this paper, will refer to natural gas (NG), biomethane, renewable natural gas (RNG), compressed natural gas (CNG), and liquid natural gas (LNG). Methane is a highly abundant hydrocarbon with the chemical formula CH_4, which can come from a variety of sources, in a variety of forms. It may contain some other chemical fuel species in minor concentrations, while remaining primarily CH_4. Methane has the distinct advantage of burning incredibly clean compared to other fuel sources. Methane systems are also remarkably similar to existing ICEs, and they can work on their own or in conjunction with other fuel sources. Methane is a very economical source of fuel. However, it requires sophisticated systems to store and distribute. Methane can be derived as a renewable source of energy and is currently being promoted as a green alternative for propulsive power in both on-road and off-road vehicles.

Utilizing CNG as a fuel source is not a new concept. Propane was utilized from the 1940s to the early 1970s as an alternative source of fuel to gasoline [40]. Minneapolis Moline produced a propane-powered tractor in 1941, and many other companies with various models followed suit for the next few decades. Propane had the advantage of being much cheaper than gasoline, and it was readily available, especially in areas of the country with large oil production. Propane tractors were inevitably phased-out due to big improvements in diesel engine technology in the 1960s and 1970s, but propane became incredibly common as a fuel for indoor applications in forklifts and other indoor ICE equipment, due to it being a much cleaner burning fuel, producing significantly less NO_x and CO. Propane-based equipment is unquestionably much safer to be around in enclosed spaces. Methane shares many of the same characteristics and is more prominent in the current era.

Methane can also be derived from a variety of sources. The most common is as a by-product of the petroleum industry. When oil is pumped from the ground, a wide variety of gaseous hydrocarbons are extracted. Crude oil is a heterogeneous mix of chemicals, consisting of chain-linked carbon atoms with hydrogen atoms bound to the free bond points. These oil molecules range from short gaseous hydrocarbons to long semi-solid tar-like hydrocarbons, with a full liquid array in between. Methane falls at the short end of the spectrum with only one carbon atom, followed by ethane, propane, and butane with 2, 3, and 4 carbons, respectively. NG contains roughly 97% methane, with the rest of the mixture being slightly larger gaseous hydrocarbons and trace amounts of other gases.

RNG is currently receiving a lot of attention as an alternative fuel source. Methane is a common by-product of many industries, such as agriculture, wastewater treatment plants, and landfills. RNG is sustainable, as it can be derived from waste products, and it contains carbon which is already circulating within the ecosystem. Industries use digesters, which utilize methane-producing microorganisms to break down organic materials. In agriculture, this is implemented within the livestock industry, particularly with dairy and swine, which produce large amounts of liquid manure slurry. In these cases, large tents are built over the storage lagoons, which capture the produced methane and other gases. Additional systems are put in place to clean and process these gases into usable RNG.

Methane is unique, because it can be used in conjunction with a variety of fuel sources and technologies. Methane-compatible engines can utilize methane from any source, which gives the operator several options, depending on what sources and technologies are available. Methane engines are like other common ICEs, such as diesel and gasoline engines. Methane needs to be ignited via another heat source or spark plug, like a gasoline engine. However, it has a high-octane value so it can run in higher compression engines, like a diesel engine. Caterpillar makes several of these models of engines [41]. Systems exist, which integrate both fuel systems together, gasoline–methane or diesel–methane. Methane-based systems hold significant potential as an alternative power source for agriculture, as the technologies currently exist and are already in mass production, making most of today's stationary installations turnkey. Mobile applications may require more thought and development.

Methane burns much cleaner than other fuels. Methane is a very simple hydrocarbon. Therefore, the combustion reaction is simple. This results in fewer unburned hydrocarbons being released into the atmosphere from an incomplete burn, which is a big problem for both gasoline and diesel engines using more complex molecules. NG is much purer than liquid fuels, which have various impurities, like sulfur, mixed-in with the fuel. These impurities create pollutants. Methane also burns cooler, so less NO_x is produced than with gasoline or diesel. The benefit of this is that these engines do not require as much air pollution control (APC) technology to meet the US or EU EPA emissions standards. APC systems, such as exhaust gas recirculation (EGR), diesel particulate filter, diesel exhaust fluid systems, among others, which are common on modern Tier IV_{Final} diesel engines, are not needed on methane-powered systems. These systems reduce the fuel efficiency of the diesel engines and can also be very problematic to operate and maintain. Additionally, they are expensive to produce and service.

The drawbacks to NG power are that the infrastructure is not nearly as widespread as it is for modern diesel fuel. While digesters provide cheap fuel, they have an expensive initial cost, which the producer will have to absorb along with the purchase of a methane-engine tractor. Most rural farm operations do not have NG lines running to their farms either. NG is stored at high pressures, which is also problematic. Fueling requires a means to transfer high pressure methane into the tractor. This requires a much more sophisticated system than a diesel fuel storage and transfer facility, and it is more dangerous for the operator. Methane storage systems are expensive, because they need to be more robust to handle the high pressures involved.

Multiple authors and companies have considered the benefits of methane-powered agricultural tractors. Bisaglia et al. (2018) conducted an analysis and research testing regarding the efficacy of utilizing methane power, and they compared it to a similar diesel tractor in side-by-side testing [42]. The test utilized a New Holland T6050® tractor outfitted with a 3 *L* Iveco methane engine [43]. The modified tractor was compared to a New Holland T6.140®, which matched its basic specifications. The engine performance and functionality in the field were monitored and found to be nominally similar. The big advantage from a methane engine would be the 85% reduction in greenhouse gases [42].

Matuszewska et al. (2016) researched the available sources of biofuels, the potential methane production from each source, and the efficiency of a dual fuel tractor running a mixture of diesel and biofuel [44]. The biggest impact of the research was that particulate and NO_x emissions were reduced compared to diesel engine operations [44]. The engine manufacturer Cummins released a report outlining the benefits and drawbacks of switching to RNG as a fuel source. The article focused on the sources of RNG and its cost savings, but in addition to the potential cost savings. The work demonstrated reduced greenhouse gas emissions from RNG fleet operations were possible [45]. CNH Industry has introduced the T7.270 LNG®, which is the first tractor run on liquid NG. Their reporting discussed the new tractor prototype and how using LNG was advantageous at increasing on-board energy density, decreasing the volume of space needed to store it in the vehicle [46]. Finally, NGV America (2022) outlined the practicality and trade-offs of transitioning a trucking fleet over to NG [47]. NGV America is a NG trade organization, and their report concentrated on the costs and the environmental impacts of a transition to RNG. They discussed the potential reduction of both traditional pollutants and greenhouse gases available from a concentrated shift from fossil fuels to RNG.

The results from scientific testing and newer technological developments provide considerable insight into the specifics of methane-based power. The main metrics worth considering are performance, economic feasibility, and vehicle range of methane-burning engines compared to diesel. In several performance measurements, the methane-powered engines proved to be equivalent to diesel engines. Unfortunately, methane is less energy dense than diesel, so it generally has a higher fuel consumption on a mass basis. However, it burns much cleaner, so the engineers can increase the fueling rate and achieve comparable engine performances. For instance, Cummins recently released the X15N® engine, which is a 15 *L* I6 engine powered by NG [48]. The engine is advertised at 400–500 *hp* and 1450–1850 $ft\text{-}lb_f$ torque. This exactly matches the standard X15 Efficiency Series® engines on a power and torque basis. However, it provides less power than the Productivity Series® and Performance Series® of the X15® engine family. The X15N® does have the advantage of being lighter, since it is not diesel powered. It is more comparable in weight and size to the X13 Series® engine. The NG engine also has the advantage of burning cleaner, and it does not require nearly as extensive an APC emissions reduction package as a diesel engine, making this engine more efficient and more reliable. Methane burns much cleaner than the larger molecules in diesel fuel and gasoline and does not create the level of unburned hydrocarbons in the exhaust that the traditional fueled engines do. This makes methane-powered engines much simpler to design for, as they do not need as many APC systems in place to capture or burn the particulate matter produced by the diesel engines. This simplification is highly advantageous, as these systems are prone to breaking and are a parasitic load on the engine.

Economically, there are many trade-offs when running a methane-powered system. Methane needs to be stored either as a CNG or as a liquid. Each of these options requires high-pressure tanks and a processing system which can compress the gas to high pressures. This additional equipment gives the NG systems a higher initial cost than the diesel storage and filling facilities. Currently, these engines are also a little more expensive to purchase, but they would be more competitive in price, if provided with an increased scale of production. For the vehicular use, the additional cost comes from having high pressure tanks and lines, but these expenses are offset by having fewer emissions systems. The price of NG per heat value released is less than diesel or other liquid fuel sources. It is estimated that with a $1.50 difference in price, a truck fleet would have a ROI of 4.76 years, with $2,500 savings in year 5, and $10,500 net savings per year for every year afterward, assuming a 7,000 *gal/mth* minimum fuel purchase [47].

The real advantage within the agricultural industry is when farms can produce their own RNG. CNH has partnered with Bennamann Ltd. to create a digester system which processes manure from cattle or swine operations to create methane that a tractor can utilize [49]. New Holland has released the T6.140 N® which runs completely on CNG, and they have released a prototype of the T7.270®, which runs off LNG. While the CNG tractor requires additional tanks, the LNG model can store energy densely enough to fit into the same volumetric footprint as the diesel model. In this case, the farmer would eliminate diesel fuel costs, after the initial capital cost of the system and any operational costs associated with running the machine were absorbed.

The feasibility of running NG is dependent on the operation. Many parts of America have access to grid-based NG or an abundant source of organic waste which can be processed. This presents an outstanding economic opportunity to capitalize on an underutilized resource, and it could prompt a growth in the NG infrastructure, which could be used beyond fueling tractors, into applications such as heating houses or running grain dryers, which currently rely on propane. Unfortunately, many places do not have this potential. Methane transportation requires trucks equipped with high-pressure tanks, which are more complicated and expensive than diesel transport tanks. This makes methane less attractive as operations become more remote from population centers. Places such as Europe, which have a lot of livestock and organic waste, as well as a denser population, would be more inclined to invest in RNG infrastructure than farming operations in West Texas or the American Great Plains.

The energy capacity of fuel is another difference between methane and diesel. Diesel fuel is liquid at room temperature and is not volatile, and it has a high energy density and can easily be stored in a tank. Methane as a gas is much less dense at atmospheric temperature and pressure. To store significant amounts, the gas needs to be highly compressed or liquified. As a gas, methane does not have an energy density comparable to diesel. The New Holland T6.140 N® has extra tanks located on the nose of the tractor and at the corners on the rear of the cab to get a comparable operational range to a diesel tractor. However, the T7.270®, which uses LNG, at a much higher energy density, is capable of storing a comparable fuel amount to diesel in the space which would have been taken up by the diesel tank and extra emissions systems, so it keeps the same volumetric footprint as the diesel model.

The tests performed in the field by scientists and manufacturers show that methane is a viable fuel source competitor to diesel. Methane-powered tractors have many applications where the technology is advantageous compared to diesel fuel. The cost savings associated with methane are a big plus for the technology. This is especially true in situations where the farm can produce the fuel. Modern digesters can provide this resource, which would have otherwise escaped into the atmosphere, and utilize it for productive work. Methane works great for a tractor or loader, doing a lot of day-to-day chores on the farm in a livestock operation, where the fuel draws are generally consistent throughout the year. Since the methane is produced at a constant rate in this situation, the infrastructure would need to be in place to ensure that the energy is either used in operation, sold to the grid, or the excess is burned in a generator to power the farm.

Methane power also proves beneficial over current diesel power in how clean the fuel combusts. Methane is a simple molecule to burn, and it reduces the risks of producing other unwanted by-products during combustion. Methane burns very clean, so there is not as large a requirement for APC systems to reduce the operational emissions. Complicated emission control systems are highly problematic in diesel engines. Propane has been used for years in indoor applications such as forklifts for this very reason. This makes the methane-powered engines more reliable in the long run, compared to modern diesel engines, since they are less complicated and require less maintenance. Methane power is much less prone to environmental regulations, so the technology may have significant longevity in places which are more inclined to strict environmental laws.

5. Biofuel propulsion

Although methane can be considered a biofuel, the biofuels in this section will focus on liquid fuels derived from plant-based sources. Plants, predominately corn and soybeans, can be utilized for production of food, fuel, and fiber. With regard to fuel, the main plant-based biofuels on the market today are ethanol and biodiesel. Ethanol is a product of corn, and biodiesel is a product of soybeans. Being plant-based, these fuels are completely renewable, unlike conventional gasoline and diesel. These fuels require combustion to release energy, which still emits pollutants and greenhouse gases. However, the amounts of emissions that ethanol and biodiesel emit are less than those of gasoline and diesel, respectively [50]. In addition to the undesired emissions, biofuels require significant manufacturing inputs at corn ethanol plants and soybean crush/refineries. Like traditional liquid fuels, both ethanol and biodiesel prices can fluctuate based on their feedstock commodity price, but overall, biofuels tend to be competitive with their traditional liquid fuel counterparts [51].

There exist many similarities between ethanol and biodiesel, along with specific advantages and drawbacks of each. First, the data surrounding ethanol and its use as a means of propulsion will be examined. According to the Office of Energy Efficiency and Renewable Energy, "..U.S. corn ethanol could produce 20% lower GHG emissions than gasoline" [52]. This clearly demonstrates that ethanol can reduce CO_2 emissions. Ethanol is very prevalent as a liquid fuel source for automotive propulsion with "more than 98% of U.S. gasoline containing some ethanol to oxygenate the fuel" [52]. The higher octane level of ethanol makes it a great candidate for blending with gasoline to make higher performance fuels. Ethanol may be cheaper than gasoline in some cases, but its price can fluctuate with corn yields, market conditions, and government subsidies. Amid its advantages, ethanol does have some strategic drawbacks. There are only limited land resources to grow corn for ethanol production, and farmland availability will only continue to decline as society develops. Additionally, ethanol is less energy dense than gasoline. As seen in **Figure 1**, ethanol's energy density is 19.9 *MJ/kg* compared to that of gasoline at 45.8 *MJ/kg* [53]. While there is an overall positive outlook for ethanol with its renewable and more environmentally friendly nature, its low energy density is one of the leading reasons why it is not more widely utilized for automotive-based liquid fueled propulsion.

Biodiesel's efficiency, advantages, and disadvantages will now be examined. Biodiesel has become increasingly popular, as the technology to craft the liquid fuel has improved. According to a US Department of Energy website, the demand for biodiesel increased by 79% from 2017 to 2022 [54]. Diesel is

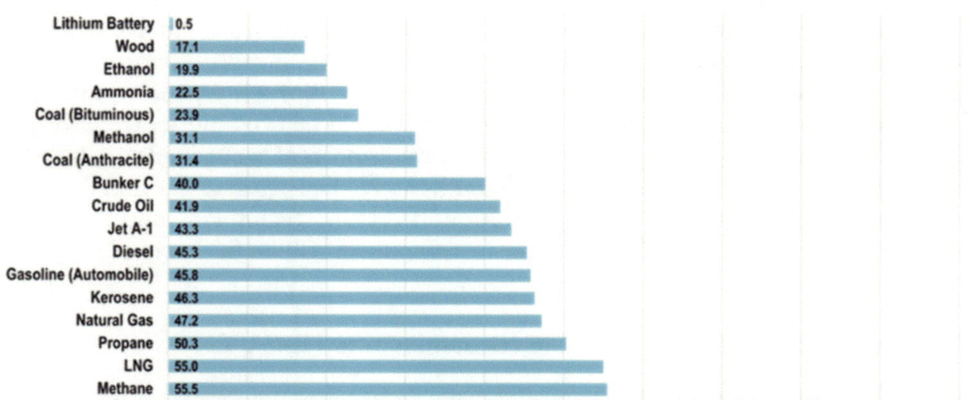

Figure 1.
Chemical energy content of some fuels in MJ/kg [51] (Rodrigue, 2024).

currently the leading propulsive fuel within the off-road vehicle industry, but biodiesel, like ethanol, is more sustainable and renewable. Biodiesel offers many of the same advantages that ethanol does, and it can be made from other resources besides soybeans, including waste cooking oils and other grown oil-grain crops. The feedstocks needed for biofuel are readily available in the United States, opening the door for less reliance on imported foreign crude oil that the production of diesel fuel requires. The downsides of biodiesel are similar to ethanol. It holds less energy per mass than diesel, but the difference is less drastic than for ethanol and gasoline. Biodiesel can be difficult to store, and it is very temperature-sensitive. Biodiesel can gel in cold temperatures, and this can add complexity and cost to its use in cold climates [55]. Biodiesel is also slightly acidic, and its use following traditional diesel within in a specific machine will require more frequent fuel filter changes as the plated contaminates from inside the fuel system are dissolved [56, 57]. Currently, both biodiesel and ethanol are blended and are not the sole fuel source for most engines. Moving forward, there are likely to be opportunities in the agricultural propulsion space to completely utilize both biodiesel and biodiesel/diesel mixtures. This will probably require the design of some modified engine technologies, as well as the production of additional crop quantities to be used for the manufacture of propulsive liquid fuels.

Biofuels have already made a dramatic impact in the aviation industry. In fact, there is a likelihood that biofuels will continue to have an even bigger impact and larger footprint within the aviation industry. Sustainable aviation fuel (SAF) is currently being heavily researched. Incentive programs exist to find smaller carbon footprints and reduced greenhouse gas emissions within aviation, in comparison to petroleum-based jet fuels. Research indicates that "aviation contributes to about 3% of the world's CO_2 emissions, which amounts to about 915 million *ton* of CO_2 per year" [58]. In addition, it is projected that "…the use of SAF will play the most important role if the sector is to achieve its 2050 decarbonization targets" [58]. The research around SAF blend stocks looks promising, not just from an environmental perspective, but also from an efficiency standpoint as well. Work by Webber at al. (2024) indicates that "…lignin conversion technologies have the near-term potential to match the enthalpic yields of existing commercial sustainable aviation fuel production processes" [59]. Lignin is an abundant renewable compound which could be very viable for energy dense SAF production. Biofuels arguably have the most potential to be utilized in the near term and could be used more extensively as an alternative fuel propulsion source. This is very important for agriculture, because jet fuel is very similar to diesel, if somewhat lighter, like kerosene. Jet fuel is produced in huge quantities to drive aviation. If the SAF technology becomes relevant for aviation, then the crossover to agriculture will be fairly simple.

6. Electric-continuously variable transmissions (E-CVTs)

A standard continuously variable transmission (CVT) provides an infinite range of ratios between the input and output shafts. These units are also known as infinitely variable transmissions (IVTs). Standard CVT and IVT units should not be confused with hybrid transmissions, which sometimes use the same branding nomenclature. An E-CVT uses electric components instead of mechanical ones to achieve the functionality. These devices allow the user to vary the gear ratio to any required value within the limits of the CVT, as compared to the 5 to 18 fixed ratios offered by a conventional manual or powershift transmission.

A mechanical CVT uses pulleys and a belt or conical rollers. The pulleys or cones in a CVT can vary their effective diameter, thus changing the ratio between input and output shafts. As the effective diameter on one shaft, usually the driver, changes, the diameter of the other adapts to accommodate and keep the distance between the shafts constant. In a belt-type device, this means that the belt continuously changes its position on the pulleys, experiencing elevated stress levels. Consequently, belts in a mechanical CVT design wear out quickly. Cone designs are also prone to accelerated wear. Engineers have tried to combat this problem with metal-core belts and advanced friction materials, but this has only slightly extended the service lives of these transmissions and makes them much more expensive to repair.

Therefore, as technology has advanced, engineers sought alternative methods of achieving CVT functionality. One of these alternatives is the E-CVT. In an E-CVT, the pulleys and belt (or cones) are replaced by a generator, motor, and power electronics controlled by a computer. The driver pulley is replaced by a generator, and the driven pulley is replaced by a motor. The power electronics between these components effectively mimic the change in effective diameter of a CVT by changing the motor's speed by varying its AC frequency. Initial E-CVTs have demonstrated much longer operational lives than mechanical CVTs.

E-CVTs have proven very popular in on-road vehicles, because varying terrain is an appropriate use of this technology. Many of the Japanese automotive manufactures have in some form or another implemented an E-CVT into their vehicles. Some examples are Toyota's Hybrid Synergy Drive®, used in the Prius® and Camry® Hybrid [60], and Nissan's e-POWER®, used in Notes® and Kicks® [61]. Honda i-MMD® is used in Accord Hybrid® and Civic Hybrid® [62].

There have also been implementations of E-CVTs in off-road vehicles. Although fewer companies have embraced E-CVTs, John Deere has a design for an E-CVT called eAutoPowr® in their 8 R series tractors [63]. Fendt uses an E-CVT called VarioDrive® in their tractors as well [64]. Manufacturers such as Komatsu have built excavators that use E-CVTs for the tracks and swing bodies [65], while Volvo has placed E-CVTs in their mining trucks [66].

Mocera and Martini (2022) conducted a compelling investigation into the potential use of hybrid E-CVTs in specialized agricultural tractors [67]. Their study explored a system where a battery supplied peak power demands, with recharging occurring during low-power periods. The results indicated that this approach could reduce diesel engine size and improve fuel savings. Sakai (1988) provided insights into early E-CVT development, considering the technological constraints of the overall design [68]. Subaru's E-CVT system, which utilized a unique magnetic powder clutch [69], was a key focus of the study. Sakai (1988) compared conventional mechanical gearboxes with E-CVTs, examining various implementations and highlighting the limitations that needed to be addressed through technological advancements [68]. Miller (2006) offered a comprehensive overview of E-CVT architecture for hybrid electric vehicles [70]. The chapter focused on establishing a classification framework for E-CVTs, analyzing their operating modes, and exploring control strategies for improved efficiency. It also discussed the advantages and disadvantages of E-CVTs. All of these authors seemed to believe that E-CVYs were extremely effective in on-road vehicles. Whether this proposition can translate into off-road use remains unproven.

An E-CVT can be divided into three major components:

- the generator;
- the electric motor; and
- power electronics.

The generator is an electrical device that converts mechanical shaft power into electrical power. Usually in E-CVTs, a separate external power source excited three-phase generator is used. The rotor creates a magnetic field that induces a current in the stator when spinning. Separately excited generators typically have a power to mass ratio of 0.5–20.0 *kW/kg* and can cost around $30,000–$50,000 for a tractor-sized generator of 500 *kW* [71].

Electric motors have made a lot of advancement over the last 20 years [72]. Today's electric motors are very power dense and usually have very high stall torques, while being able to rotate at very high speeds. **Table 4** provides the details on the three of the most popular types of electric motor types, showing that a permanent magnet synchronous motor (PMSM) has the highest power density. This type of motor uses rare earth materials and can be very expensive to manufacture. Hence, a lot of research is being done to improve the power density of traditional induction motors (IMs) and switch reluctance motors (SRMs) [73]. SRMs have high potential and can provide very high 0 *rpm* torque, as well as act as regenerative brakes.

The power electronics are the most crucial part of an E-CVT. Gallium nitride (GaN) semiconductors are generally used for power electronics circuits, because they have higher breakdown voltages. This means that the E-CVTs can operate at higher voltages, thus reducing the current supplied to the electric motor at the required speed. There have been a lot of advancements in the field of power electronics over the last 20 years, specifically in the field of semiconductors [74]. The

Motor type	Typical applications	Power density (*kW/kg*)	Advantages	Disadvantages	Cost ($/*W*)
PMSM	Passenger cars, hybrid vehicles	2.0–4.0	High efficiency, compact size, wide operating range	Can be more expensive than IMs	0.5–1.5
IM	Industrial equipment, some electric vehicles	1.0–3.0	Robust, reliable, cost-effective	Lower efficiency than PMSMs	0.2–0.5
SRM	Electric vehicles, heavy-duty applications	1.0–2.5	Simple construction, fault tolerant	Higher torque ripple, acoustic noise	0.3–1.0

Table 4.
Comparison of electric motor types [72, 73].

latest GaN semiconductors can handle more power than the standard silicon semiconductors, and they offer 40% more thermal conductivity. By reducing the current needed for operation, GaN semiconductors can significantly reduce the amount of copper wiring needed within the drive system. There have also been advancements in power control systems within the software algorithms. They can automatically optimize their amplification gains for the current operational situation and the hardware being utilized, reducing operational costs [74]. All of these advancements mean that today's power electronics can consistently achieve greater than 98% operational efficiencies.

An E-CVT can further be classified based on the topology of the specific design being used. Power split devices, such as the motor generator units (MGUs) as shown in **Figure 2**, are connected to the sun and the ring gears of a planetary gearbox, while the engine is connected to the carrier. Optionally, a battery may be used to provide extra power where required or for recovering power from regenerative braking.

In the series E-CVT type, the engine is completely disconnected from the driveline, with only a generator unit (GU) connected to it. A MGU is connected to the final drive/differential, and the connection between the GU and the MGU is made by power electronics. Optionally, a battery can be connected with the power electronics to make this a hybrid system. These types of systems are most often used in diesel–electric locomotives. A series E-CVT is less efficient than a power split unit. **Figure 3** provides a schematic of the two E-CVT styles side-by-side for comparison.

There are multiple advantages accrued when using an E-CVT within a vehicle. Optimized engine operation improves fuel economy, because E-CVTs have infinite gearing ratios. They allow an engine to operate at the most optimal and efficient point on its load map. This means that the engine stays within the valley of the most efficient BSFC, thus saving fuel while providing optimal power for the vehicle's mission. Regenerative braking is also possible with a hybrid vehicle. Energy can be

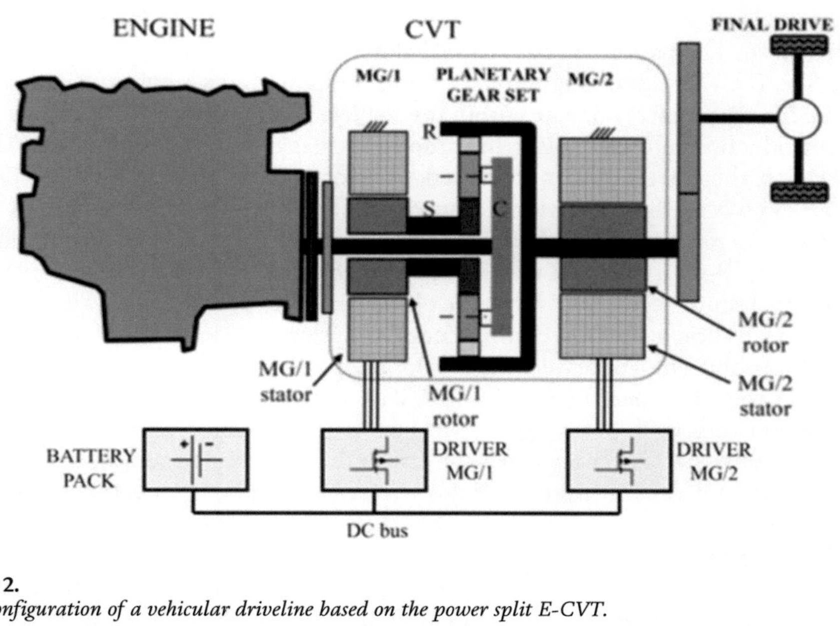

Figure 2.
Basic configuration of a vehicular driveline based on the power split E-CVT.

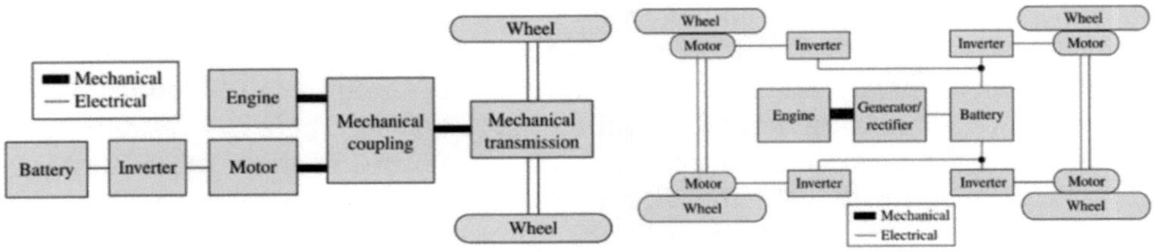

Figure 3.
A comparison of the two basic topologies of vehicular driveline based on series E-CVT, with a power split E-CVT on the left and a series E-CVT on the right.

recovered during the braking process by using the motor connected to the wheels as a generator and charging the batteries. This will reduce the strain on the engine and improve efficiency, particularly for on-road vehicles.

E-CVTs enhance the performance and productivity of vehicles. They provide precise speed control, because E-CVTs have infinite gear ratios. Within limits, they can achieve any ground speed required, without altering the speed of the engine. This is helpful in an off-road vehicle, as the engine is also used to drive hydraulics pumps and other components for ancillary activities, which may not be dependent on vehicle speed. Electric motors increase torque at low speeds, compared to ICEs. They can output extremely high torques at low speeds. This improves traction and pulling power. E-CVTs provide smooth operation. Because the gear ratio is continuously variable, speed changes up or down happen continuously, without the jumps commonly felt when using a convectional transmission.

E-CVTs can reduce emissions from a vehicle by allowing the use of a smaller engine. E-CVTs enable the reduction of engine weight and physical size within a specific vehicle design volume, and since the engines do not need to adhere to the ratio is given by the conventional transmission, there are no holes in the power curve for the vehicle either. The engine size reduction can be enhanced further, if a hybrid system is used. The potential of additional electrification is simpler using an E-CVT design. The E-CVTs essentially have the same components as a hybrid EV, so the potential for regenerative power within the vehicle is increased. E-CVTs allow for an easier introduction of overall electrification to the basic overall design of the vehicle.

However, there are significant challenges associated with using E-CVTs. E-CVTs are more expensive than traditional transmissions. There are more advanced technologies involved, and the base components are more expensive than traditional transmissions. E-CVTs require specialized components, like power electronics and a cooling system to reject heat from the electric motor, power electronics, and batteries in a hybrid system. A complex control system is also required to operate the E-CVT smoothly. This increases the development costs, engineering, and the component costs for specific vehicles. E-CVTs do not handle high peak power requirements very well. Many agricultural tasks are strenuous and have elevated high-power duty cycles. These tasks can strain the electrical system in an E-CVT, specifically the power control electronics, which can lead to reduced performance and a shorter life. Additionally, off-road hybrid vehicles that use E-CVTs have extended duty cycles that can deplete the battery. This will lead to reduced performance, unless the work being done is paused, and the batteries are charged. This is extremely unlikely to happen in agricultural use. Off-road vehicles are typically operated in harsh conditions and under heavy duty cycles. This means that the components of the E-CTV must be rated and certified to work under these conditions, which can make them bulkier and more expensive to manufacture. The maintenance cycles for off-road agricultural vehicles are performed less frequently, as downtime is lost productivity and revenue to farmers. This means that the components must be overbuilt to handle the workload without failing, again, causing additional upfront expenses.

Although these challenges might seem to preclude the use of E-CVTs in agricultural vehicles, the potential benefits are too large to ignore. There is active research going on to reduce the cost of the system by removing rare earth materials, as well as to increase the durability of the overall assemblies by reducing the number of components utilized. Since the concept of using E-CVTs in off-road vehicles was first proposed, many advancements in their designs have been made, especially in the field of power electronics. These components are now more efficient, compact, and cheaper than previously thought possible. If these trends continue, the positive potential within E-CVTs may eventually overcome the apparent shortcomings of use within the agricultural market.

7. Hydrostatic transmission

A hydrostatic transmission (HST) is a type of transmission that uses hydraulic oil as its working fluid. It consists of two major parts: a hydraulic pump that is driven by the engine and a hydraulic motor that drives the wheels or tracks. HSTs are very popular on construction machinery, as these vehicles are already primarily powered by hydraulics. The pump and the motor are usually directly connected to each other to offer the best efficiency possible. However, there are some cases where there are valves that are used as throttling elements to control the speed of the motor.

In general, HSTs can be grouped into five major categories. A fixed displacement pump and fixed displacement motor is known as a Type I device and is shown in **Figure 4**. These are the easiest and cheapest type of HST to both conceptualize and build. The pump and the motor displacements are both fixed. This also means that there is a direct coupling and a single ratio for the transmission. This ratio can be selected by choosing different displacements for the pump and the motor. The speed of the engine must be varied to change the speed of the output shaft, because of the fixed ratio. This negates all the usefulness of this transmission type, and thus these units are rarely used.

A variable displacement pump and fixed displacement motor is known as a Type II device and is shown in **Figure 5**. In this type of HST, the displacement of the pump is

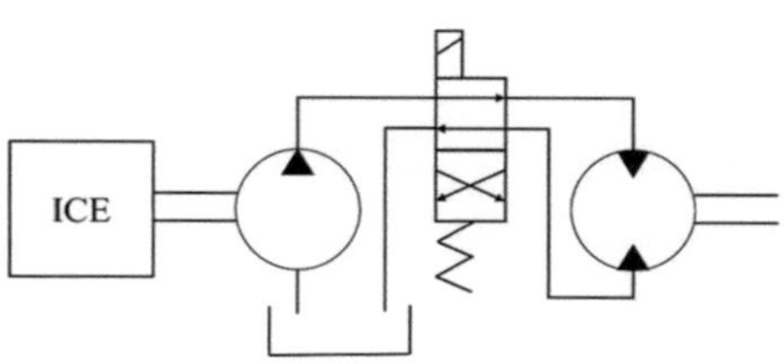

Figure 4.
Basic design of a Type I fixed displacement pump and fixed displacement motor HST.

variable, and the displacement of the motor is fixed. The transmission ratio can be varied by varying the displacement of the pump. By selecting a pump that can reverse flow directions, the transmission can vary its output shaft rotation into the reverse direction. This type of transmission usually uses a variable displacement swash plate type of piston pump. These units are typically limited to low load applications, like lawn mowers, small tractors, forklifts, or other small and light duty equipment.

A fixed displacement pump and variable displacement motor is known as a Type III device and is shown in **Figure 6**. In this type of transmission, the displacement of the pump is fixed, and the displacement of the motor is variable. The ratio of this transmission can be varied, but instead of providing a linear and smooth change of direction, these units create a large jump in speed when changing directions. This is generally undesirable, and hence, the Type III unit is seldom used in mobile applications. The major advantage of this type of HST is that multiple motors can be driven by a single pump, and the speeds of the individual motors can be varied independently at each driven point. Therefore, these devices have found a niche in conveyor belts and hydraulic-powered devices within industrial plant settings.

A variable displacement pump and variable displacement motor is known as a Type IV device and is shown in **Figure 7**. In this type of HST, the displacements of

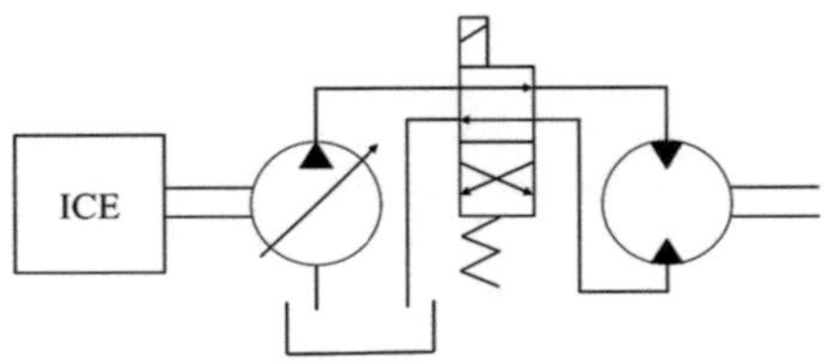

Figure 5.
Basic design of a Type II variable displacement pump and fixed displacement motor HST.

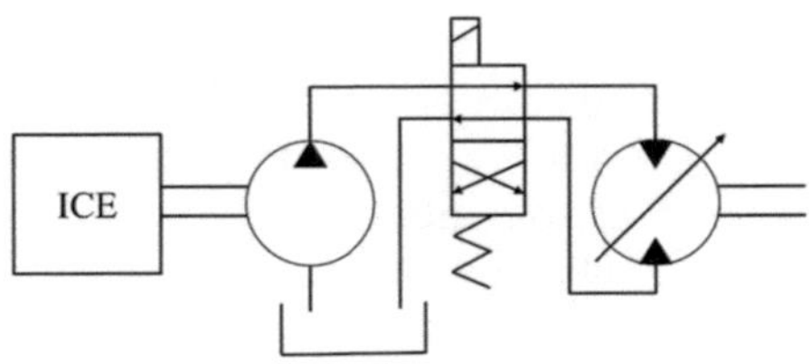

Figure 6.
Basic design of a Type III fixed displacement pump and variable displacement motor HST.

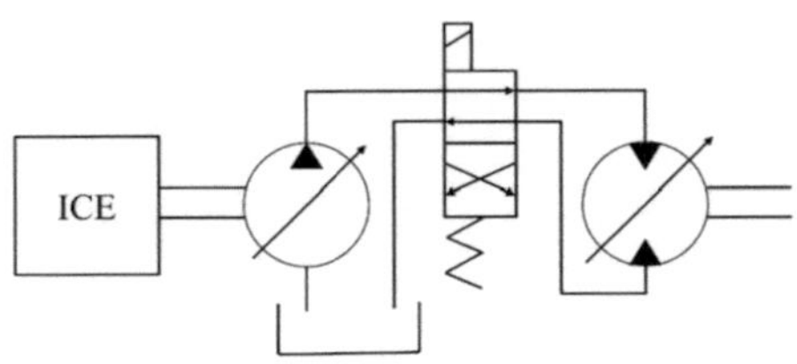

Figure 7.
Basic design of a Type IV variable displacement pump and variable displacement motor HST.

both the pump and the motor are variable. These units offer an infinite variability of the transmission ratio, and they can be coupled with other types of transmissions. Coupled transmissions are usually considered to be power split HSTs. Because these units have two variable displacement hydraulic components, they are more complex, expensive, and less efficient than other types of HSTs available on the market. The major advantage of the Type IV HST is that they are robust and can handle very large loads. They are therefore the preferred type of HST for construction and agricultural machinery.

Another type of transmission which uses a HST is the hybrid transmission, sometimes called a power split system. Hybrid transmissions use both hydrostatic and mechanical sections. The mechanical section provides efficiency, and the hydrostatic portion of the transmission gives it infinite speed ratio variability. They come in two types, defined by their hard coupling point in the assembly. A forward or input coupled hybrid transmission is shown in **Figure 8**, and a rearward or output coupled unit is shown in **Figure 9**. Both of these assemblies offer significant fuel savings to the operator by moving most of the power through the mechanical transmission and using the hydrostatic section to provide the exact speed match for the operation, but due to its efficiency at lower speeds, the rearward coupled unit is preferable for off-road applications. The input coupled device would perform slightly better at on-road uses. This allows the engine to operate at peak BSFC, and the vehicle speed to be controlled independently. These transmissions have significant potential in propulsive agricultural tractors and are currently available from all major tractor manufacturers.

The entire field of HST is an extremely active research area, with several groups pursuing improved HST designs and efficiencies. Kugi et al. (2000) focused on the modeling of variable displacement HSTs and control strategies for variable displacement HST [75]. This chapter showed significant advancement in detailed

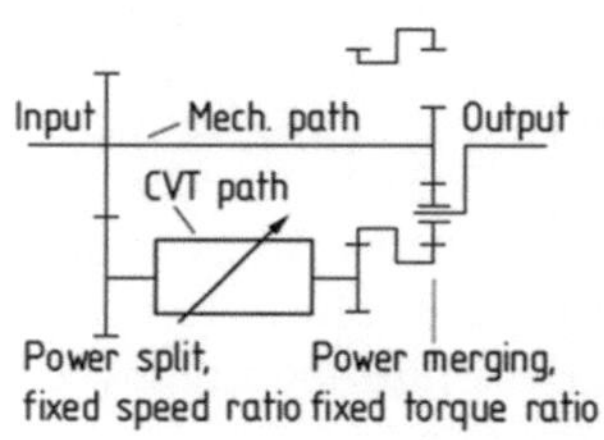

Figure 8.
Basic design of a forward (input) coupled hybrid transmission.

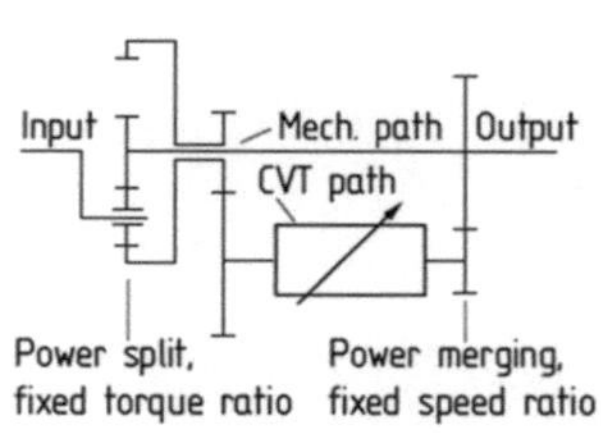

Figure 9.
Basic design of a rearward (output) coupled hybrid transmission.

component modeling and the simulation of HSTs, showcasing their potentials and shortcomings. Manring and Luecke (1998) discussed the modeling and simulation of fixed displacement HSTs and greatly improved the state-of-the-art in simulation models by considering dynamic effects of output loadings [76]. This research conducted experiments to calibrate the simulation model and validate the results obtained by the model. Ho and Ahn (2010) focused on the energy losses in the HSTs and showed how these losses could be recovered, especially in scenarios where energy regeneration was possible [77]. Detailed modeling was conducted, and energy regeneration strategies were considered, after simulation indicating the most promising strategies was conducted. Active HST research centers with ongoing state-of-the-art work are present in multiple companies and universities.

Hydrostatic transmissions are mostly used in vehicles where a direct shaft connection between the power source and the tractive drive is not possible or is inconvenient. Construction machines and modern agricultural combines are typically propelled using hydraulic transmissions. Most machines that use traditional HSTs operate using a Type II or Type IV design topology. The other two architectures don't offer enough advantages to be of use in vehicular applications. Many of the large tractors using HSTs today are provided with the hybrid transmission style unit. This enables the engine RPM to remain constant, while offering IVT ratios. Modest utility tractors and construction equipment are typically provided with Type IV transmissions to increase their versatility and value on the job sites. A variable displacement pump with a fixed displacement motor is a popular choice for smaller, less expensive vehicles, as these units are cheaper to manufacture than the Type IV architecture.

When looking at Type III transmissions with fixed pump and variable displacement motor designs, as shown in **Figure 10**, there are two basic variants which can be used: a directional control valve (DCV) type and a four-quadrant motor type. The first architecture uses a DCV to control the direction of rotation of the motor, which requires a two-quadrant variable displacement hydraulic motor for the output shaft. In the second variant, the DCV is omitted by using a four-quadrant hydraulic pump/motor at the output shaft. The first design is less expensive, but it comes with the disadvantage that if the DCV is switched abruptly during operation, an over-pressure situation can occur. A shock loading from the over-pressure is then applied to both the hydraulic and mechanical components of the system, which can damage the drivetrain. Hence, a safety interlock must be employed to stop the DCV from changing at high speeds or high displacement settings in the components. The second topology has the advantage that a shock loading of the hydraulic and the mechanical system is

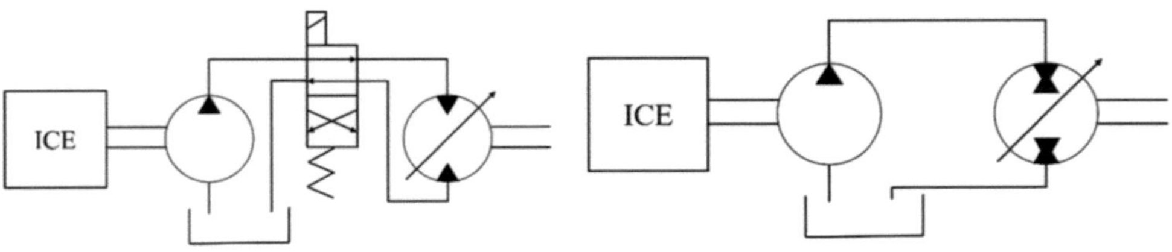

Figure 10.
Two basic variants for Type III fixed pump and variable displacement motor HSTs, with a DCV type on the left and a four-quadrant motor type on the right.

far less likely to occur, but four-quadrant hydraulic machines that can have over-center capabilities are complex and expensive to install. These machines also are less efficient and noisier than variable displacement two-quadrant hydraulic motors.

When looking at the Type IV variable pump and variable motor hydraulic transmission topologies, a similar split in topology can be observed. A similar analysis can be performed with these configurations, and the end result is similar. Designs with DCVs are less expensive to build, while designs with four-quadrant hydraulic motors are more expensive to operate. The additional expense does not generally provide benefits worth the extra capital cost upfront, unless installed on an expensive machine. Larger, more costly construction equipment typically use the second topology with the four-quadrant hydraulic pump motor, whereas smaller construction machines and tractors use the design with the DCV. Since most off-road machines come to a stop before changing direction, the topology with the DCV poses a smaller risk.

A HST efficiency can be further improved by adapting it into a hybrid style configuration and installing an accumulator in the hydraulic circuit. This will enable regenerative braking, where the energy from braking is stored in the accumulator and then fed back into the system for further use, upon restarting. This is a simple and often used strategy, since the basic principles of hydraulic systems make energy recovery fairly simple.

There are numerous advantages and challenges encountered when using HST units in agricultural machinery [78]. HSTs offer very smooth and continuous control of transmission ratios. This is possible across a wide range of speeds, but performance at both the low and high ends of the transmission's range can entail problems. By selecting a low displacement for the pump and a high displacement for the motor, a HST can considerably multiply the torque applied to the output shaft. This is especially beneficial for construction machines. HSTs are compact and simple, because hydraulics are very power dense. HTSs generally contain fewer components, making them low maintenance and easy to repair. Finally, by combining HSTs with accumulators, energy recovery is possible, and thus, the efficiency of HST units can be improved through the addition of optional components.

Unfortunately, HSTs, where both the pump and the motor have variable displacements, are often inefficient and have a lot of internal power losses. Since hydraulic machines have very high tolerances and are difficult to manufacture, the costs for components can be fairly high, especially with variable displacement machines. The pumps and motors can be very expensive to replace. The control of the displacement of both the pumps and motors can prove challenging at times, and thus, they require complex controllers for optimal operation. This also increases the cost to assemble these machines. Finally, hydraulic oil in HSTs needs to be continually filtered and changed, due to the degradation of oil from continuous heating cycles and from metal contamination due to wearing components in the systems. The required continual maintenance can increase the downtime and costs associated with HSTs.

8. Conclusions and recommendations

This chapter has presented an evaluation of the potential of electric, diesel–electric hybrid, methane and NG, hydrogen, and biofuel as possible alternative power sources for agricultural vehicles. E-CVTs and HSTs were examined as alternative drivetrains for improving efficiencies in agricultural tractors. Additionally, each system had its own unique characteristics, including its advantages and drawbacks, presented. Outside of specialized in-the-barn use, all-electric tractors are probably multiple decades away for typical general agricultural use, and then, they are only likely if significant advances in battery technology become available. On the other hand, diesel–electric hybrids represent a feasible and positive potential technology that could be employed today under the proper economic circumstances. Hydrogen-based ICE engines could also be utilized today, if there were a hydrogen-based infrastructure present to support the equipment. Hydrogen fuel cell technology would have the same difficulty in rural locations, but it also needs further technology development to become robust enough to survive agricultural use. Hydrogen technology, while feasible, ranks as near term, at best in availability, due to the supply and transportation infrastructure deficiencies. Methane-based propulsion is feasible currently, under the right circumstances. If an agricultural operation can generate its own fuel, this power system is economical and environmentally advantageous today. Biofuels are also currently technically feasible, but they depend upon government subsidies to have any economic viability. Rising general crude oil costs could significantly change the picture for biofuels, but for now this technology must also be classified as near term. More development work within all of these alternative fuel fields will undoubtedly improve the chances for their deployment within agriculture.

E-CVT technology is a promising energy efficiency technology, but its lack of an experience base within heavy duty applications will limit its use in agriculture, until its durability can be demonstrated. Without this experience base in heavy equipment, E-CVTs must be classified as near term. HSTs currently suffer from efficiency problems, limiting their use as propulsive transmissions in agricultural applications requiring drawbar power. They work well in nonpulling types of applications. Hybrid transmissions are currently demonstrating their economic benefits in agricultural tractors, and this should only continue to improve moving forward. The fuel savings potential from these components will drive further acceptance in the agricultural markets. Ongoing research and development will continue to improve these existing technologies for agricultural power units and create new options beyond the direct replacements for diesel-fueled machines described here. These will undoubtedly provide unique and positive benefits to the industry. The pursuit of potential alternatives to petroleum fuels is ongoing and will likely require multiple solutions and constant improvement, due to the variety of scenarios present within the greater realm of the industrial-age agricultural world. Hopefully, this review has provided a comprehensive examination of the available technologies for off-road alternative power sources and drivetrains which can be utilized by engineers and technicians within the agricultural, electrical, and mechanical fields to improve the energy pathways within these critical pieces of machinery and advance the state-of-the-art for this equipment into a renewable and sustainable future.

Acknowledgments

The Fall 2024 Design of Off-Road Vehicles class at Purdue University's School of Agricultural and Biological Engineering is acknowledged for their contributions to the structure and content of this technical chapter. The USDA/NIFA provided partial support for this effort through proposal #2023-11313, 'The Societal Implications of Clean Energy Tractors for Farming'. Some of the figures and tables within this chapter are from referenced sources, who have provided their consent for inclusion within this document. Dr. Carol S. Stwalley is graciously thanked for her editorial work with the manuscript and in the final formatting of the document. An Artificial Intelligence (AI)-powered large language model software, ChatGPT, by Microsoft, Incorporated (Redmond, Washington) was used to aid in the organization of this effort, but all manuscript elements are original from the authors. Any errors are entirely the result of the authors. The mention of trade names or commercial products in this article is solely for the purpose of providing specific technical information and does not imply recommendation or endorsement by Purdue University. The findings and conclusions in this publication are those of the authors alone, and they should not be construed to represent any official Purdue University determination or policy. Purdue University is an equal opportunity/equal access organization.

Author details

Elijah Collins, Reece Durbin, Nicholas Hoff, Brandon Kitch, Porter Little, Tovia Owens, Seshan Calapatti Suresh and Robert M. Stwalley III*
Purdue University Agricultural and Biological Engineering, USA

*Address all correspondence to: rms3@purdue.edu

References

[1] Williams M, "Machinery Milestones: Electric tractor power." 2019. Available: https://www.fwi.co.uk/machinery/technology/machinery-milestones-electric-tractor-power [Accessed: 2024-December-5]

[2] Perner A, Vetter J. Lithium-ion batteries for hybrid electric vehicles and battery electric vehicles. Advances in Battery Technologies for Electric Vehicles. 2015;173–190. DOI: 10.1016/B978-1-78242-377-5.00008-X

[3] Nelson P, Amine K, Rousseau A, "Advanced Lithium-Ion batteries for plug-in hybrid-electric vehicles," Argonne National Laboratory, Lemont, IL, 2007. https://www.researchgate.net/publication/228732600_Advanced_Lithium-Ion_batteries_for_plug-in_hybrid-electric_vehicles [Accessed: 2024-December-5]

[4] Nitta N. "Electric Tractors are Rolling Out in the Field. Here's What That Could Mean for Farmers." 2023. Available at: https://modernfarmer.com/2023/04/electric-tractors-are-rolling-out-in-the-field-heres-what-that-could-mean-for-farmers/ [Accessed: 2024-December-5]

[5] Zimeno Inc. dba Monarch Tractor, "Met the Mk-V Tractor". [Online]. 2024. Available: https://www.monarchtractor.com/contact-us [Accessed: 2024-December-5]

[6] CNH Industrial N.V., "T4 Electric Power," 2024. [Online]. Available: https://agriculture.newholland.com/en-gb/europe/products/agricultural-tractors/t4-electric-power [Accessed: 2024-December-5]

[7] Zimeno Inc. dba Monarch Tractor, "Electric Tractors in Farming: 15 Benefits," 2024. [Online]. Available: https://www.monarchtractor.com/blog/electric-tractors-farming [Accessed: 2024-December-5]

[8] Baek S-Y, Jeon -H-H, Kim W-S, Kim Y-S, Park C-G, Kim Y-J. Data analysis and traction performance evaluation of an electric all-wheel drive tractor during agricultural operation. Journal of the ASABE. 2024;**67**(5):1217–1229. DOI: 10.13031/ja.15952

[9] Deng X, Sun H, Lu Z, Cheng Z, An Y, Chen H. Research on dynamic analysis and experimental study of the distributed drive electric tractor. Agriculture. 2023;**13**(1):40. DOI: 10.3390/agriculture13010040

[10] Kernan P, Font E, Keeling J, "Pacific Northwest Electric Tractors Barrier Study," 2022. [Online]. Available: https://www.cadeogroup.com/wp-content/uploads/2022/04/PNW-Tractor-Electrification-Study.pdf [Accessed: 2024-December-5]

[11] European Inland Barging Innovation Platform (EIBIP), Hybrid/Diesel-Electric Propulsion. [Online]. 2018. Available: https://eibip.eu/publication/hybrid-diesel-electric-propulsion/ [Accessed: 2024-December-5]

[12] International Council on Clean Transportation, "Global adoption trends of hybrid propulsion systems in commercial vehicles (2024-03)," 2024. [Online]. Available: https://www.icct.org/publications/hybrid-adoption-2024 [Accessed: 2024-December-5]

[13] Elkafas AG, Shouman MR. A study of the performance of ship diesel-electric propulsion systems from an environmental, energy efficiency, and economic perspective. Marine Technology Society Journal. 2022;**56** (1):52–58. DOI: 10.4031/MTSJ.56.1.3

[14] International Council on Clean Transportation. "Incentives for electrifying agricultural tractors in India (5Oct22)". [Online]. 2022. Available: https://theicct.org/publication/india-hvs-evs-incentives-elec-tractors-india-oct22/ [Accessed: 2024-December-5]

[15] Union Pacific Railroad Company. "Diesel-Electric Locomotives". [Online]. 2024. Available: https://www.up.com/aboutup/special_trains/diesel-electric/&lang=en [Accessed: 2024-December-5]

[16] Lu X, Wu Y, Lian J, Zhang Y, Chen C, Wang P, Meng P. Energy management of hybrid electric vehicles: A review of energy optimization of fuel cell hybrid power system based on genetic algorithm. Energy Conversion and Management. 2020;**205**:112474. DOI: 10.1016/j.enconman.2020.112474

[17] General Dynamics Land Systems Division. General Dynamics Corporation, "General Dynamics business units to participate in AUSA 2022". [Online]. 2022. Available: https://www.gdls.com/general-dynamics-business-units-to-participate-in-ausa-2022/ [Accessed: 2024-December-5]

[18] Trevithick J. "Next Generation AbramsX Tank Will Have Hybrid Power Plant". [Online]. 2022. Available: https://www.twz.com/next-generation-abramsx-tank-will-have-hybrid-powerplant [Accessed: 2024-December-5]

[19] Lajunen A, Kivekäs K, Freyermuth V, Vijayagopal R, Kim N. Simulation-Based Assessment of Energy Consumption of Alternative Powertrains in Agricultural Tractors. World Electric Vehicle Journal. 2024;**15**(3):86. DOI: 10.3390/wevj15030086

[20] Cunanan CJ, Tran M-K, Lee Y, Kwok S, Leung V, Fowler M. A review of heavy-duty vehicle powertrain technologies: Diesel engine Vehicles, battery electric vehicles, and hydrogen fuel cell electric vehicles. Lean Technologies. 2021;**3**(2):474–489. DOI: 10.3390/cleantechnol3020028

[21] Bunton J, Casterton D, Czub J, Downey A, Haslett N, Kinnaman W, McIntire J, Lengacher J, Stwalley III RM. Electrical Hybridization of Agricultural Power Units. In: Electric and Hybrid Vehicles - New Perspectives on Technologies and Control Challenges. London, UK: InTech Open; 2025

[22] Nebergall J, "Hydrogen engine insights for truck and bus manufacturers," 2022. [Online]. Available: https://selfscreening.cummins.com/en-in/news/2022/06/13/hydrogen-engine-insights-truck-and-bus-manufacturers [Accessed: 2024-December-5]

[23] National Aeronautics and Space Administration (NASA), "Apollo experience report: Power generation system". 1973. [Online]. Available: https://ntrs.nasa.gov/citations/19730011146 [Accessed: 2024-December-5]

[24] Bockris JO. The Hydrogen economy. Environmental Chemistry. 1977;549–582. DOI: 10.1007/978-1-4615-6921-3_17

[25] Lanz A, Heffel J, Messer J. "Hydrogen Fuel Cell Engines and Related Technologies". [Online]. 2001. Available: https://www.energy.gov/sites/default/files/2014/03/f9/fcm00r0.pdf [Accessed: 2024-December-5]

[26] Airships.net, "The Hindenburg Disaster," 2017. [Online]. Available: https://www.airships.net/hindenburg/disaster/ [Accessed: 2024-December-5]

[27] Reclamebureau A. "H2 Dual Power". [Online]. 2024. Available: https://h2dualpower.com/en [Accessed: 2024-December-5]

[28] First Mode Holdings, Inc. "Cleaning up the future of mining. [Online]. 2021. Available: https://firstmode.com/updates/case_studies/clean-energy/#:~:text=For%20the%20initial%20prototype%2C%20First%20Mode%E2%80%99s%20plan%20was,use%20when%20bursts%20of%20extra%20power%20are%20needed [Accessed: 2024-December-5]

[29] A.G D. "Deutz TCG 7.8 H2 – Hydrogen engine. [Online]. 2024. Available: https://www.deutz.com/en/products/hydrogen-engines/ [Accessed: 2024-December-5]

[30] Cummins, Inc.. "Hydrogen engines can help you reduce emissions". [Online]. 2024. Available: https://www.cummins.com/engines/hydrogen [Accessed: 2024-December-5]

[31] Caterpillar, Inc.. "Renewable Hydrogen Power Solutions". [Online]. 2024. Available: https://www.cat.com/en_US/by-industry/electric-power/electric-power-industries/hydrogen.html [Accessed: 2024-December-5]

[32] CNH Industrial N.V.. "Steyr® and Tu Wien Unveil FCTRAC Biogenic Hydrogen–Powered Tractor Project". [Online]. 2024. Available: https://media.cnh.com/EMEA/steyr/steyr–and-tu-wien-unveil-fctrac-biogenic-hydrogen-powered-tractor-project/s/558423b9-c9ca-4ca6-a201-b592ff8cf154 [Accessed: 2024-December-5]

[33] Gomes Antunes JM, Mikalsen R, Roskilly AP. An experimental study of a direct injection compression ignition hydrogen engine. International Journal of Hydrogen Energy. 2009;**34**(15):6516–6522. DOI: 10.1016/j.ijhydene.2009.05.142

[34] Gomes Antunes JM, Mikalsen R, Roskilly AP. An investigation of hydrogen-fuelled HCCI engine performance and operation. International Journal of Hydrogen Energy. 2008;**33**(20):5823–5828. DOI: 10.1016/j.ijhydene.2008.07.121

[35] Sandalci T, Karagöz Y. Experimental investigation of the combustion characteristics, emissions and performance of hydrogen port fuel injection in a diesel engine. International Journal of Hydrogen Energy. 2014;**39**(32):18480–18489. DOI: 10.1016/j.ijhydene.2014.09.044

[36] Bose P, Banerjee R, Deb M. Effect of hydrogen-diesel combustion on the performance and combustion parameters of a dual fueled diesel engine. International Journal of Energy and Environment. 2013;**4**(3):497–510

[37] Senthil Kumar M, Ramesh A, Nagalingam B. Use of hydrogen to enhance the performance of a vegetable oil fuelled compression ignition engine. International Journal of Hydrogen Energy. 2003;**28**(10):1143–1154. DOI: 10.1016/S0360-3199(02)00234-3

[38] Naganuma K, Takagi Y, Kawamura A, Sato Y. Study of NOx emissions reduction strategy for a naturally aspirated 4-cylinder direct injection hydrogen ICE. SAE Technical Papers. 2010;2010–01–2163. DOI: 10.4271/2010-01-2163

[39] Villenave N, Dayma G, Brequigny P, Foucher F. Experimental measurements of ultra-lean hydrogen ignition delays using a rapid compression machine under internal combustion engine conditions. Fuel. 2024;**355**:129431. DOI: 10.1016/j.fuel.2023.129431

[40] Mowitz D. "Propane-powered tractors: A Bygone Era". [Online]. 2023. Available: https://www.agriculture.com/propane-powered-tractors-a-bygone-era-8414824 [Accessed: 2024-December-5]

[41] Caterpillar, Inc.. "CAT® G3500 Engines for Gas Compression. [Online]. 2024. Available: https://www.cat.com/en_US/by-industry/oil-and-gas/gas-compression/g3500-engines.html [Accessed: 2024-December-5]

[42] Bisaglia C, Brambilla M, Cutini M, Fiorati S, Howell M. Methane/Gasoline Bi-Fuel engines as a power source for standard agriculture tractors: Development and testing activities. Applied Engineering in Agriculture. 2018;**34**(2):365–375. DOI: 10.13031/aea.12262

[43] S.v.a I. "Natural Gas Powered Vehicles". [Online]. 2024. Available: https://www.iveco.com/en-us/press-room/kit/pages/e3-naturalgaspoweredvehicles.aspx [Accessed: 2024-December-5]

[44] Matuszewska A, Owczuk M, Zamojska-Jaroszewicz A, Jakubiak-Lasocka J, Lasocki L, Orlinski P. Evaluation of the biological methane potential of various feedstock for the production of biogas to supply agricultural tractors. Energy Conversion and Management. 2016;**125**:309–319. DOI: 10.1016/J.ENCONMAN.2016.02.072

[45] Cummins, Inc.. "Cummins Natural Gas Engines". [Online]. 2022. Available: https://www.cummins.com/sites/default/files/2022-12/cummins-natural-gas-engines-ebook.pdf [Accessed: 2024-December-5]

[46] CNH Industrial N.V. New Holland Agriculture debuts world's first LNG tractor". [Online]. 2022. Available: https://media.cnh.com/emea/cnh/new-holland-agriculture-debuts-world-s-first-lng-tractor/s/68636ec4-451c-4e7b-8826-4e08d794dc68 [Accessed: 2024-December-5]

[47] Natural Gas Vehicles for America (NGV America), "Start now. RNG is how.," 2022. [Online]. Available: https://transportproject.org/wp-content/uploads/2022/05/Start-Now-RNG-is-How-Report-5.2.2022.pdf [Accessed: 2024-December-5]

[48] Cummins, Inc., "X15N (2024)," 2024. [Online]. Available: https://www.cummins.com/engines/x15n-2024 [Accessed: 2024-December-5]

[49] CNH Industrial N.V., "CNH Industrial takes controlling stake in Bennamann," 2023. [Online]. Available: https://media.cnh.com/emea/bigbaler/cnh-industrial-takes-controlling-stake-in-bennamann/s/aa67280a-fcd1-44df-8f89-4b56bffb81cb [Accessed: 2024-December-5]

[50] Hofer FJ, Arnberger A, Kruth R, "Viability of Alternative Fuels to

Decarbonize the World's Largest Agricultural Tractor Market," in *2024 Symposium on International Automotive Technology*, Warrendale, PA, 2024. doi:10.4271/2024-26-0065

[51] Resendez J, "Biodiesel vs Diesel: Understanding the Key Differences," 2023. [Online]. Available: https://www.biodieselfoundation.org/biodiesel-vs-diesel/ [Accessed: 2024-December-5]

[52] Sarisky-Reed V, "Ethanol vs. Petroleum-Based Fuel Carbon Emissions," 2022. [Online]. Available: https://www.energy.gov/eere/bioenergy/articles/ethanol-vs-petroleum-based-fuel-carbon-emissions [Accessed: 2024-December-5]

[53] Rodrigue J-P, "Energy Density of some Combustibles," 2024. [Online]. Available: https://transportgeography.org/contents/chapter4/transportation-and-energy/combustibles-energy-content/ [Accessed: 2024-December-5]

[54] US Department of Energy Office of Energy Efficiency and Renewable Energy, "Biodiesel Production and Distribution," 2024. [Online]. Available: https://afdc.energy.gov/fuels/biodiesel-production [Accessed: 2024-December-5]

[55] Resendez J, "Maximizing Diesel Shelf Life: Key Strategies," 2023. [Online]. Available: https://www.biodieselfoundation.org/how-long-diesel-can-be-stored/ [Accessed: 2024-December-5]

[56] Goering CE, Hansen AC. Engine and Tractor Power, St. Joseph, MI: American Society of Agricultural & Biological Engineering (ASABE; 2014

[57] Goering CE, Stone ML, Smith DW, Turnquist PK. Off-Road Vehicle Engineering Principles, St. Joseph, MI: American Society of Agricultural Engineers (ASAE; 2003

[58] van Dyk S, Saddler J, "Progress in Commercialization of Biojet/Sustainable Aviation Fuels (SAF): Technologies and Policies," 2024. [Online]. Available: https://task39.ieabioenergy.com/wp-content/uploads/sites/37/2024/05/IEA-Bioenergy-Task-39-SAF-report.pdf [Accessed: 2024-December-5]

[59] Webber MS, Watson J, Zhu J, Jang JH, Caglayan M, Heyne JS, Beckham GT, Roman-Leshkov Y. Lignin deoxygenation for the production of sustainable aviation fuel blendstocks. Nature Materials. 2024;**23**(12):1622–1638. DOI: 10.1038/s41563-024-02024-6

[60] Corporation TM, "Toyota's Develops New Hybrid System," 2003. [Online]. Available: https://global.toyota/en/detail/220017 [Accessed: 2024-December-5]

[61] Corporation NM, "e-POWER system," 2024. [Online]. Available: https://www.nissan-global.com/EN/INNOVATION/TECHNOLOGY/ARCHIVE/E_POWER/ [Accessed: 2024-December-5]

[62] Honda Motor Company, "Honda Hybrid Technology The Power of Two: I-MMD," 2024. [Online]. Available: https://www.honda.co.uk/cars/world-of-honda/present/hybrid/performance.html [Accessed: 2024-December-5]

[63] Deere & Company. John Deere 8R 410 now available with eAutoPowr. [Online]. 2022. Available: https://www.deere.co.uk/en-gb/our-company/news/john-deere-8r-410-now-available-with-

eautopowr–210334 [Accessed: 2024-December-5]

[64] Agco GmbH. “VarioDrive”. [Online]. 2024. Available: https://www.fendt.com/int/agricultural-machinery/tractors/variodrive [Accessed: 2024-December-5]

[65] Komatsu Europe International N. V.. “Komatsu Hybrid System for excavators”. [Online]. 2008. Available: https://www.komatsu.eu/en/komatsu-hybrid-technology [Accessed: 2024-December-5]

[66] Volvo AB. “Volvo FE. Your Versatile City Truck”. [Online]. 2024. Available: https://www.volvotrucks.com/en-en/trucks/models/volvo-fe.html [Accessed: 2024-December-5]

[67] Mocera F, Martini V. Numerical performance investigation of a hybrid eCVT specialized Agricultural Tractor. Applied Sciences. 2022;**12**(5):2438. DOI: 10.3390/app12052438

[68] Sakai Y. The “ECVT” Electro Continuously Variable Transmission. In: 1988 SAE International Congress and Exposition. Warrendale, PA: 1988. DOI: 10.4271/880481

[69] Hartig J, “eCVT transmission – What Is It? How Does It Work?,” 2024. [Online]. Available: https://roadsumo.com/ecvt-transmission/ [Accessed: 2024-December-5]

[70] Miller JM. Hybrid electric vehicle propulsion system architectures of the e-CVT type. IEEE Transactions on Power Electronics. 2006;**21**(3):756–767. DOI: 10.1109/TPEL.2006.872372

[71] Bonauer S. Is Replacing a CVT Transmission Worth It? 7 Factors to Consider. [Online] 2024. Available: https://www.cardeeply.com/is-replacing-a-cvt-transmission-worth-it/ [Accessed: 2024-December-5]

[72] Mair A. Maximizing Efficiency: The Next Frontier in Electric Motor Technology. [Online]. 2024. Available: https://www.powermotiontech.com/technologies/motors/article/55093806/maximizing-efficiency-the-next-frontier-in-electric-motor-technology [Accessed: 2024-December-5]

[73] Hua W, Zhao G, Hua H, Cheng M. General power equation of switched reluctance machines and power density comparison. *IEEE Transactions on Industry Applications*. 2017. pp. 1–1. DOI: 10.1109/TIA.2017.2700476

[74] Duvall R. “What’s Next for Power Electronics? Beyond Silicon”. [Online]. 2024. Available: https://www.keysight.com/blogs/en/tech/sim-des/eda/pe/what-s-next-for-power-electronics-beyond-silicon [Accessed: 2024-December-5]

[75] Kugi A, Schlacher K, Aitzetmüller H, Hirmann G. Modeling and simulation of a hydrostatic transmission with variable-displacement pump. Mathematics and Computers in Simulation. 2000;**43**(4-6). DOI: 10.1016/S0378-4754(00)00234-2

[76] Manring ND, Luecke GR. Modeling and designing a hydrostatic transmission with a fixed-displacement motor. ASME Journal of Dynamic Systems, Measurement, and Control. 1998;**120**(1):45–49. DOI: 10.1115/1.2801320

[77] Ho TH, Ahn KK. Modeling and simulation of hydrostatic transmission system with energy regeneration using hydraulic accumulator. Journal of

Mechanical Science and Technology. 2010;**24**:1163–1175. DOI: 10.1007/s12206-010-0313-8

[78] Comellas C, Pijuan J, Potau X, Nogués M, Roca J. Efficiency sensitivity analysis of a hydrostatic transmission for an off-road multiple axle vehicle. International Journal of Automotive Technology. 2013;**14**:151–161. DOI: 10.1007/s12239-013-0017-z

Section 2

Energy Engineering for Batteries

Chapter 6

Computational Modelling of Electrode/Electrolyte Interfaces: Challenges and Perspectives from Li-Ion to Post Li-Ion Batteries

Hepsibah Priyadarshini Christopher and Sudha Venkatachalam

Abstract

This chapter presents a detailed review of the computational methodologies employed in order to characterise the electrode/electrolyte interface of Post Li-ion batteries (LiBs) in conjunction with the approaches developed for Li-ion battery technology. Insights on the parameters that must be evaluated for materials to function as effective electrode and electrolyte candidates in conjunction with microstructural as well as heterogeneous effects at the interface are discussed with inputs from various models documented in the literature. Besides, information pertaining to the role of defects (Schottky, Frenkel, etc.), grains, grain boundaries, anions, cations and additives is presented to provide the reader a holistic view on multifarious interfacial phenomena. Moreover, density functional theory (DFT), Grand Canonical Nudged Elastic Band (GCNEB), ab initio molecular dynamics (AIMD), hybrid MD based on ReaxFF parameterisation and machine learning (ML) approaches that shed light on diverse electrochemical interfacial reactions are highlighted. In addition, the challenges involved in computational modelling as well as methodologies that can be incorporated to overrule the associated bottlenecks are furnished. Thus, this chapter will be extremely instrumental for experimentalists and theoreticians in designing better electrode and electrolyte candidates utilising abundant, non-toxic and cost-effective materials which will guarantee an enhancement in energy density, voltage, capacity and cycling stability.

Keywords: electrode/electrolyte interface, density functional theory, grand canonical nudged elastic band, hybrid MD, heterogeneous effects, machine learning

1. Introduction

In general, employment of fossil fuels to power automobiles causes three crucial issues: oil demand, global warming and air pollution. Therefore, rechargeable batteries have become the front-runners in catering to the energy demands of human community by advancing towards a future free of fossil fuels [1]. In that, Li-ion chemistry is known for its attractive attributes including enhanced energy density, low or no memory effect as well as lesser self-discharge [1, 2]. Although, it exhibits the aforementioned features, the global demand for Li is anticipated to surge to 200,000–300,000 per year, and the currently available resources can last approximately for 90 years [3]. This necessitates the design of new electrode and electrolyte candidates that are abundant, affordable and non-toxic while ensuring an increase in energy density, capacity and voltage. Hence, a lot of research has been devoted towards the exploration of alternate battery chemistries including Na, Mg, Ca, Al, K and Zn.

Moreover, the liquid electrolytes utilised in batteries succumb to various drawbacks including flammability and leakage which pose serious hazards such as fire and explosion [4, 5]. Thus, solid electrolytes (SEs) that are safer and promise superior battery performance have been receiving enhanced attention [6]. Further, the solid-state batteries (SSBs) that are the research challenge of the 2020s are anticipated to be the coveted technology of the 2030s. Though the characteristics of individual materials influence the battery efficiency, the interfaces within greatly determine the success of SSBs. The prominent interfacial issues associated with solid-state electrolytes (SSEs) are ion immobilisation, creation of space charge layers, growth of dendrites, microstructure (defects, grain boundary, etc.), lattice mismatch, thermal instabilities, reactivity at the solid electrolyte interphase (SEI) and maintenance of contact between solid electrolyte and electrode materials [7]. Most importantly, literature revolving around the computational understanding of both the heterogeneous and microstructural interface of SEs is scarce. This might be due to the computational complexity involved in simulating the most interactive region comprising of defects, grains, grain boundaries, anions, cations and additives. In addition, studies on the modelling of the electrode/electrolyte interface may give critical insights into the interfacial resistance, which is the biggest threat to the performance of the SSBs. However, detailed interfacial studies with respect to Li chemistry are available, whereas with other post-Li-ion chemistries, it is almost negligible.

Thus, this chapter is structured as follows: Firstly, theoretical methodologies applied to gauge the suitability of electrode candidates have been detailed. In the second part, the characteristics that an SSE ought to possess from a computational viewpoint are presented. Thirdly, the theoretical approaches employed to scrutinise the interface have been discussed. Furthermore, the challenges in modelling the interfacial reactions and the strategies that can be adopted to model them with good accuracy have been furnished.

2. Theoretical approaches to gauge the performance of electrode

The following subsections detail the various density functional theory (DFT) and ab initio molecular dynamics (AIMD) techniques employed to characterise the electrode performance in terms of their structural stability, reaction voltage, electronic properties, ionic conductivity, diffusivity and adsorption kinetics.

2.1 Key considerations for ensuring structural stability

In general, the cyclability and electrochemical performance of the prospective electrode candidates are largely determined by their structural stability. The evaluation of phonon frequencies, formation, cohesive and Gibbs free energies is vital for ascertaining the stability of the structure [8]. Besides, the cohesive and formation energies derived from DFT calculations by plotting against the respective volume in addition to fitting the values in accordance with the Birch Murnaghan equation was highly effective in predicting the stability of novel candidates [9]. Considering Li_2MnSiO_4 polymorphs, the *Pmnb* phase emerged to be the energetically plausible one from energy vs. volume (E-V) curve and this finding was found to be consistent with the experimental results [10]. Interestingly, subjecting this material to high pressure under experimental conditions yielded the theoretically reported second most stable phase, $Pmn2_1$ [10]. Whilst cohesive and formation energies are relevant in determining the stability of multifarious materials at 0 K, computing the Gibbs free energies serves to be highly efficient for calculating similar properties under different temperatures and pressures. The Gibbs free energies of 14 structural modes of Na_2FeSiO_4 have been calculated by Zhu et al. by employing DFT methods [11] and the breakdown of the results is as follows. The *Pn* phase appeared to be the dominant one when both pressure and temperature were maintained at lower values. Conversely, escalations in temperature to 500 K effectuated the transformation towards the $Pna2_1$ phase. In addition, scaling up the pressure to 8 GPa facilitated its transition to $Pca2_1$ polymorph. Therefore, DFT studies that furnish crucial information by assessing the free energies of challenging polymorphisms confer additional insights to the experimentalists. In order to understand the stability of electrode materials from the lattice dynamics perspective, the information gained from phonon dispersion curves becomes mandatory. However, these studies are computationally tedious and can only be applied to scrutinise the dynamic stability of systems comprising fewer atoms. In view of the phonon spectra of phosphorous containing anodes such as Na_5P_4 and Li_4P_3, the lack of imaginary frequencies in the first Brillouin zone decisively demonstrated their exceeding stability [12].

2.2 Determination of reaction voltage

Secondly, the material of choice should exhibit a higher cell voltage since it has a direct influence on the specific energy and energy density of the battery, which are both significant in escalating its performance. In essence, the voltage is determined from the energy difference between the intercalated and deintercalated fractions *via* DFT calculations. Also, the estimation of thermodynamically favourable intermediate phases across different compositions furnishes accurate approximations to the voltage curve. Primarily, the voltage assessed *via* the Generalised Gradient Approximation (GGA) method was found to record values that differed by an order of 1 eV when compared to the experimentally documented ones [13]. This discrepancy has been linked to the self-interaction error, which is responsible for the underestimation in the prediction of voltage and forbidden energy gap values. Subsequently, the application of Hubbard corrections to highly correlated systems eliminated this error and resulted in voltage values that were consistent with experimental findings [14]. Later, the inclusion of temperature parameter in the cluster expansion (CE) technique within the Monte Carlo framework for Na_2MnO_2 yielded voltage and formation energy values which exhibited lesser deviations from those that were derived experimentally [15]. Nevertheless, voltage profiles obtained *via* the CE method confer valuable information on the evaluation of energy density. Furthermore, Arroyo-deDompablo et al. [16] explored the perovskites of $CaMO_3$ type as cathodic candidates for calcium ion batteries (CiBs) where M corresponds to transition metals (TM) including Cr, Mo,

Ni, Fe, Mn and Co *via* computational and experimental routes. Amongst all the TMs considered, Fe and Co-based perovskites exhibited voltage values in the desired 3–4 V range. In addition, the electrolytes tailored for CiBs perform well within the potential window of 1–4 V and hence the screened perovskites demonstrated their capacity in acting as efficient cathodes for CiBs. Thus, calculating the reaction voltage is directly linked in determining the usability of the electrode material in addition to aid the computation of energy density.

2.3 Prediction of electronic structure

Since the electronic structure studies are extremely instrumental in assessing the battery performance, DFT studies that provide explicit perception into the hybridisation, bonding nature, conductivity, electron localisation and bond strength are accounted here. In solid-state physics, the band structure unravels information pertaining to the accessibility of available electronic states within the electrode material, which can be linked to its conductivity. In the case of Mg_3N_2, though the band structure exhibited the semiconductor behaviour with a band gap of 1.92 eV, the lithiated material behaved like a metallic one and hence demonstrated remarkable rate capacity owing to its outstanding electronic conductivity [17]. Meanwhile, the projected density of states (PDOS) was plotted and the conductivity corresponding to each desodiation fraction for Na_2MnSiO_4 was evaluated by Sudarsanan et al. [18]. A significant decrease in the forbidden energy gap values from 2.4 eV in Na_2MnSiO_4 to 0.7 eV was observed in $MnSiO_4$ and this was accredited to the intermediate band formed between the orbitals of Mn 3d and O 2p. Also, an identical energy range was observed for the up-spin states of both Mn 3d and O 2p orbitals within the valence band, corroborating the generation of a directional covalent bond. Hence, PDOS studies serve as an excellent tool in accounting for the bonding nature as well as hybridisation.

Moreover, Dixit et al. [19] scrutinised the effect of Al^{3+} doping in the NCM 523 lattice and unravelled its chemical bonding aspects including bond strength and nature through Integrated Crystal Orbital Hamilton Population (ICOHP) analysis. The Al^{3+}–O^{2-} bond in Al-doped NCM 523 resulted in an escalated ICOHP value of −2.55/bond, showcasing the presence of a strong ionocovalent bond. Dua et al. [20] analysed the electron localisation function (ELF) to investigate the interaction existing between twin graphene and adsorbed Mg atoms. The prevalence of ionic bonding between Mg and the graphene substrate was confirmed from the red regions on the ELF plot. In addition, the blue zones on the plot revealed the presence of carbon-carbon covalent bonds in the substrate. Thus, ELF is a useful tool to predict the interactions prevailing between diverse atoms in the electrode material of interest. While displaying charge density clouds and chemical environments of prospective electrode candidates *via* experimental approaches present significant challenges, charge density difference plots (CDDP) derived from DFT studies stand out as a powerful tool for visualising electron density clouds. On investigating the CDDP in SnO_2-doped $MoSe_2$ electrode for sodium ion batteries (SiBs), Zhao et al. [21] found the transfer of charges from surface Se in $MoSe_2$ to O in SnO_2, which was revealed by a loss of charges on the Se atoms, thereby confirming the chemical bonding persisting between Se and O atoms. From these observations, they had confirmed that the doping of SnO_2 has bestowed $MoSe_2$ with enhanced conductivity, thereby enabling the lamellar surface to exhibit a high-rate capacity that is significant for the effective functioning of SiBs. In addition, Wu et al. [22] explored the characteristics of $V_2O_5{\cdot}H_2O$ as host materials for Zn^{2+} ions in aqueous zinc ion batteries (ZiBs). Computation of Bader charges on the individual atoms demonstrated conclusively that a substantial amount of charge donated by Zn was procured by V and O atoms of the electrode material. Besides, the bonded water molecule was established to play an important role in accelerating the electrochemical performance of V_2O_5. In order to obtain an accurate picture of the charges on individual atoms as well as the charge transferred between various atoms, both Bader charges and CDDP provide invaluable insights.

2.4 Nudged elastic band

In addition to these, the execution of nudged elastic band (NEB) simulation unravels the estimation of key parameters such as activation energy and ionic conductivity. These simulations help in the identification of activation energy barriers and transition states that are often challenging to obtain through experimental techniques. For instance, NEB simulation of $LiTiS_2$ indicated that hopping of Li^+ ions in a divacancy and triple vacancy required less energy compared to an isolated vacancy [23]. Moreover, these studies shed light on the variation of the energy barrier as a function of the ion migration direction. In the case of Li diffusion in $LiFe(SO_4)F$ along the (101) direction, it yielded a barrier energy of 700 meV, whereas for the (111) plane it required an energy of 208 meV [24]. Similarly, the diffusion behaviour of Ca^{2+} ions in three perovskites including $Ca_2Fe_2O_5$, $CaMn_4O_8$ and $Ca_2Mn_2O_5$ was analysed from their respective energy barriers by Arroyo-deDompablo et al. [25]. In $CaMn_4O_8$, it was found that the diffusion energy barrier of Ca^{2+} ion along all the pathways exceeded 1.5 eV, and it was attributed to the presence of triangular face tunnels, which hindered Ca migration. In contrast to this, the migration of Li^+ and Mg^{2+} ions along the same pathway across these tunnels required very low activation energies. Therefore, the requirement for higher energy by Ca^{2+} ions was accredited to the synergetic effect of the larger size and charge. Whereas in $Ca_2Mn_2O_5$, the channel shares four of its edges with the neighbouring Mn polyhedra, escalating the energy barrier to 1.8 eV owing to the existence of electrostatic repulsion between the TM atoms. Amongst the various pathways considered for $Ca_2Fe_2O_5$, the transition state where Ca does not share faces with Fe polyhedra in addition to the nearest Fe atoms located at distances of 3 Å substantially reduced the electrostatic repulsions, thereby facilitating the migration of the Ca^{2+} ion with the lowest energy barrier. Importantly, lower activation energy barriers correlate with higher rate capabilities, enabling ions to diffuse in and out of the host material more efficiently. Therefore, these computational studies are vital for screening innovative battery materials, as electrode materials requiring lower activation energy showcase enhanced electrochemical activity.

2.5 Molecular dynamics simulation

Ab initio molecular dynamics (AIMD) simulation is a powerful tool for the accurate evaluation of ion transport kinetics as well as diffusion properties by integrating Newton's equation of motion over time. In contrast to NEB simulation, MD allows for the incorporation of parameters including temperature and pressure, which enables description of the system as a function of time, which is comparable to the experimental scenario. The variable charge potential employed by Ma et al. [26] for MD simulation in FeF_2 and FeF_3 aided the precise evaluation of elastic constants, surface energies, angular as well as pair distribution functions. Furthermore, the concomitant intercalation and conversion reactions with the onset of Li in both FeF_2 and FeF_3 had been showcased with remarkable precision. Besides, the (001) surface which facilitated faster diffusion of Li^+ ions as obtained from the NEB studies was selected as the substrate for cognizing both the rapid and slow diffusion rates. Considering the former case, 72 Li^+ ions were introduced every 20 ps employing the ReaxFF approach. Although the intercalation process was found to be limited, Fe atoms were reduced to their metallic states thereby generating LiF, a product of conversion reaction. In the latter case, where 9 Li^+ ions were added every 10 ps exhibited the feasibility of insertion reaction up to five layers. However, beyond these layers the structure underwent amorphisation with LiF embedded in a metallic matrix of Fe atoms. Thus, through this elaborate study, the importance of crystal surface orientation in tandem with structural evolution of the system as a function of time has been highlighted by Ma et al. [27] with absolute precision. Moreover, Su et al. [28] revealed the commencement of the conversion reaction between Li and NiO through MD studies. Besides, an increment in Li–O contacts with a concurrent decrease in the number of Ni–O bonds corroborated the generation of Li_2O, thereby revealing the onset of the conversion reaction. Consequently, we have carried out Car-Parinello Molecular Dynamics simulation (CPMD) to unravel the mechanism of lithiation in the 2 × 2 × 3 surface of Ag_2O at 298 K employing the NVT ensemble. Here, we found that the number of Ag–O (Li–Ag) decreased from 71 to 66 (16 to 1) as a function of time, whereas there was a steady presence of Li–O bonds from 0.6 ps [29], as seen in **Figure 1**. Hence, this confirms the predominance of the conversion mechanism rather than compound formation. Interestingly, we observed that Li acts as an antisite defect inducing guest for Ag in the host Ag_2O. Hence, these studies indicate the advancement of NEB calculation in the estimation of electrode performance as a precursor to the post-Li-ion battery materials.

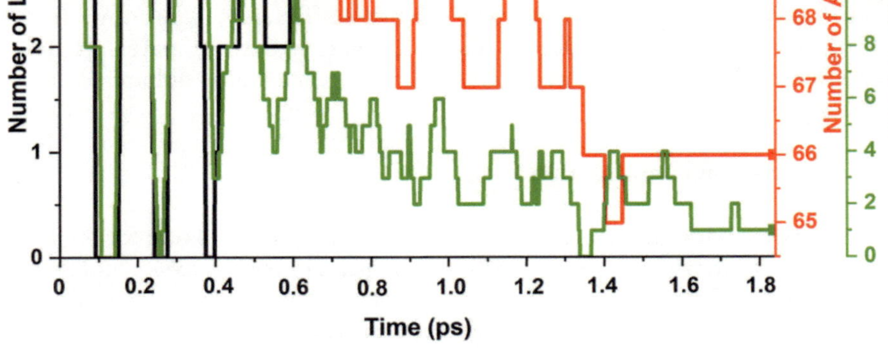

Figure 1.
Variation in the number of Li–Ag, Ag–O and Li–O bonds with respect to time of the NVT run during MD simulation [29].

DOI: http://dx.doi.org/10.5772/intechopen.1009755

2.6 Adsorption kinetics

In general, ion adsorption influences the reactivity of the electrode candidates, thereby impacting the capacity and cycling performance of the batteries. Moreover, the evaluation of adsorption energies facilitates the elucidation of atomistic interactions persisting between metal ions and the surface of interest. Kiran et al. [30] have scrutinised the adsorption of Li in Cu_2O through DFT calculations. Here, Li relaxed at the interstitial site of Cu_2O, which has been used for further studies owing to its exorbitant stability. The adsorption of a single Li in the Cu_2O lattice displayed the presence of strong Cu-Li contacts in all the investigated structures, showing the predominance of the alloying mechanism. Meanwhile, increments in the concentration of Li to two exhibited, and there is an enlargement in Cu–O bonds with the concurrent stabilisation of Li–O bonds, thereby hinting at a strong probability for the onset of the conversion mechanism. Likewise, we have conducted a systematic study by probing the adsorption of Li and Na atoms on the 2 × 2 × 3 supercell of Ag_2O [29, 31] and have drawn the following conclusions: (i) the predominance of sodiation over lithiation was observed from the higher negative binding energies observed for Na adsorption. (ii) Adsorption of single atom resulted in compound formation for both Li. (iii) Increments in the concentration of Li favoured the conversion mechanism by generating Li_2O, whereas for Na, the compound formation, which might give rise to the alloying mechanism, was observed. Overall, the variations in the ionic radius of Li and Na ions have been accredited to the observed difference in mechanisms. Therefore, determining the mechanism associated with various metal ions is essential since it sheds light on both the reversibility and capacity of batteries.

3. Theoretical studies to scrutinise electrolytes for boosting battery performance

The characteristics that a solid electrolyte ought to possess for efficient functioning of the battery, as estimated from theoretical studies, are presented in the following sections.

3.1 Estimation of thermodynamic and electrochemical stability

In essence, the thermodynamic stability of materials is evaluated based on two conditions, namely, $E_f < 0$ and $\Delta E_h = 0$, where E_f and ΔE_h represent the formation enthalpy and convex hull distance, which can be evaluated from Gibbs free energies at various temperatures and pressures. The negative values of formation enthalpy indicate the stability of the considered compound. On the other hand, the distance to the convex hull, ΔE_h is utilised in predicting the ability of the compound to undergo decomposition into other phases. When the compounds record zero or very small values (<50 meV) for ΔE_h, they are considered to be thermodynamically stable. Based on these criteria, Singh et al [32] explored the stability of 630 existing and novel antiperovskites (APs), and amongst them, 22 of the novel candidates emerged with exceeding stability. Besides, 36 compounds with ΔE_h values between 50 and 100 meV were procured, which might be synthesised under non-equilibrium conditions. Most of the experimentally synthesised APs were found to fully adhere to the criteria, with few exceptional candidates exhibiting non-collinear magnetism owing to their geometrical frustration.

Hence, the estimation of these energies serves as a significant tool to predict the thermodynamic stability of SEs, which has not been explored so far.

Also, the approach to evaluate the electrochemical stability of several SEs has been demonstrated efficiently by Schwietert et al. [6]. Here, the electrochemical windows are classified into two, one is based on the stability of degradation products (decomposition window), and the other is the indirect degradation of intercalated/deintercalated ions (intrinsic window). In the case of sulphide-based electrolytes, it has been noticed that the intrinsic window is larger than the decomposition window. In general, the intrinsic window of an SE is determined using DFT calculations based on variations in formation energies during the insertion and extraction of metal ions. Moreover, the intrinsic window for $Li_{10}GeP_2S_{12}$ (LGPS) lying between 1.19 and 2.38 V, as obtained from DFT calculations, is in very good agreement with the experimentally reported electrochemical stability window (ESW) of 1.2–2.5 V. Hence, in order to design novel solid-state batteries, experimentalists require a comprehensive understanding of the realistic ESW of solid electrolytes, which can be provided by theoretical calculations.

3.2 Evaluation of mechanical and dynamical stability

Generally, the calculation of elastic constants provides detailed information about surfaces under elastic distortions, thereby shedding light on their mechanical stability. Besides, it serves as a fundamental parameter for designing SEs that are compatible with the anodic and cathodic compartments. Singh et al. [32] have evaluated the mechanical stability of 58 cubic APs by employing DFT calculations. Mostly, cubic systems possess three distinct elastic constants: C11, C12 and C44. Furthermore, materials with cubic symmetry must satisfy the following elastic constant criteria, viz., $C11 > 0$, $C11 - C12 > 0$, $C11 + 2C12 > 0$ and $C44 > 0$. The evaluation of elastic constants for most of the APs demonstrated excellent mechanical stability by adhering to this criterion, whereas some, including Mn_3CuN, Mn_3InN and Mn_3InC, deviated from the aforementioned rule. In addition, it was observed that all the novel compounds that exhibited excellent thermodynamical stability were also mechanically stable. Meanwhile, understanding the isotropic nature of SEs is essential for developing improved electrolyte materials. In terms of mechanical properties such as bulk and Young's moduli, a spherical distribution indicates that these properties remain uniform in all directions, a characteristic known as isotropy. For instance, in $Li_7La_3Zr_2O_{12}$, the bulk modulus is significantly higher along the (001) direction compared to the (100) and (010) planes, demonstrating anisotropic behaviour [33]. Similarly, studies of Young's modulus in Li_3PO_4 show strong anisotropy along the (100) direction [33]. Also, these studies reveal that the selection of isotropic solid electrolytes plays a crucial role in improving the durability, efficiency and longevity of the devices. Therefore, isotropic behaviour is desired for SEs, and hence, the exploration of SEs that showcase spherical distributions with respect to bulk and Young's moduli should be explored further.

The dynamical stability of SEs can be achieved by analysing their phonon dispersion spectra. In the case of Li_3OCl, the spectra indicated imaginary frequencies at the M and R points at 0 K, thereby indicating its instability. Meanwhile, the same compound has been reported to be stable at 298 K from experimental studies. In order to obtain the corresponding trajectories, Chuang et al. [34] carried out AIMD studies and consequently corrected the interatomic force constants. Thus, the renormalised phonon spectra lacked negative frequencies, which is in accordance with the experimental findings. Thus, investigating the phonon dispersion by correcting the force constants and exploring the anharmonicity becomes very crucial in gaining knowledge on the dynamic stability of AP-based solid electrolytes.

3.3 Electronic conductivity measurements

With respect to the electronic conductivity, the prospective electrolyte candidate should have a lower value to minimise the onset of decomposition reactions. This can very well be determined from the electronic structure calculations using the DFT approach. On the other hand, high electronic conductivity, accelerating dendrite formation, has been exclusively accounted for by Han et al. [35]. Similarly, Marana et al. [36] have estimated the electronic structure of Li rich SE, β-Li_3S_4, as an insulator with an indirect band gap of 4.75 eV. Also, they had observed various surfaces of β-Li_3S_4 and concluded that the terminated structure along (100) and (210) directions were the most stable ones, favouring Li^+ ion mobility with higher band gap values of 4.66 and 4.20 eV, respectively. Moreover, the evaluation of cluster anions containing APs (Li_3OBH_4, Li_3OBF_4, Li_3OAlH_4, etc.) by Fang et al. [37] has led to significant conclusions that these larger ions offer higher band gaps as well as larger channels that accelerate a conducive environment for the diffusion of Li^+ ions. Further, examination of the Na-rich APs of type Na_3SA, where A site corresponded to cluster anions like ICl_4, IO_4, ClO_4 and AlF_4 yielded a wider forbidden energy gap and higher ionic conductivity values [37]. Thus, the substitution of cluster anions at the A site rather than other sites has been established to promote lower electronic conductivity. Hence, investigating surface terminations in addition to probing the substitution of ions at the desired sites will result in wider gaps, which is essential for achieving higher ionic mobilities.

3.4 Determination of ionic conductivity

Escalated ionic conductivity in SEs is one of the most coveted properties since it promotes efficient ion transport, suppresses dendrite formation and reduces internal resistance, thereby ensuring stable electrochemical performance. Yu et al. [38] have identified a series of Li and Na mixed anti-Ruddleson Popper (RP) phase APs with the formula, $Na_{4-c}Li_cAX_4$. AIMD simulation at 298 K for the optimised $Na_3LiS_{0.5}O_{0.5}I_2$ structure revealed lower activation energy and higher ionic conductivity values of 0.12 eV and 6.3 mS/cm, respectively. Moreover, Fang et al. [39] reported the superionic conductivity of 10^{-2} S/cm in Li rich APs such as Li_3SBF_4 and $Li_3S(BF_4)_{0.5}Cl_{0.5}$ by performing MD simulation at room temperature. Interestingly, Tsai et al. [40] demonstrated the excellent Na ion conductivity in double cluster anion AP, $Na_{3-x}O_{1-x}(NH_2)_x(BH_4)$, *via* AIMD. Moreover, these studies highlighted the translational and rotational dynamics of cation and anion, respectively. Motional constraints were applied to NH_2^- and BH_4^- ions where one of them remained frozen in order to estimate the interdepending behaviour of the others. On freezing the BH_4^- anions, the rotational motion showcased by the amide ion decreased considerably with absolute quenching of the translational movements of the Na^+ ion. Although the freezing of amide ions did not bring about a restriction in the anionic rotation of BH_4, it still inhibited the translation of Na cation. Thus, it became conspicuous that the rotational movement of both the anions (double paddlewheel effect) is mandatory to ensure the movement of the Na ion, which directly corresponds to the conductivity.

3.5 Scrutinising the influence of defects

It has been corroborated by MD simulation that perfect crystalline structures result in poor ionic conductivity, whereas the introduction of defects enhances the ion migration process. In the case of Li_3OCl, Lü et al. [41] have identified three types of defects including Schottky defect pair (LiCl and Li_2O) and Li interstitial with a substitution of O at the Cl site. Amongst these, MD studies revealed that LiCl deficient Li_3OCl is the most favourable structure owing to the lowest diffusion coefficient of Li ions, which is consistent with that of the experimental values. Kim et al. [42] have unravelled the performance of multivalent Mg and Ca-based APs by subjecting the materials of 10 different compositions to structural optimisation and thereafter estimated their defect formation energies amongst various other properties. They have modelled a variety of defects such as Frenkel, Schottky, neutral and charged vacancy/interstitial ones. Besides, it has been found that the defect formation energies fall in the range of 1.2–3.45 eV for all the defects. Thus, it has been concluded that higher values of ionic conductivity are achievable by regulating the defect concentration via doping. Chao et al. [43] modelled a single Li defect in Li_3ScCl_6 and Li_3InCl_6 and found enhanced diffusion behaviour of Li^+ ions, whereas uneven defect distribution at higher concentrations led to restrictions in mobility of Li ions. Hence, care should be taken when creating the defective site as well as its concentration.

4. Theoretical approaches applied for evaluating properties at the electrode/electrolyte interface

The plethora of computational studies carried out so far has been devoted to the estimation of solid electrolyte characteristics at the bulk scale. Besides, the electrode/electrolyte interface is one of the least researched subjects owing to the computational complexity involved. However, computational studies accounting for the interface of LiBs from microstructure and heterostructure viewpoints are discussed below.

4.1 Characterisation of grain boundaries

Grain boundaries can be defined as the contact surfaces between differently oriented crystal planes, and their structural features vary significantly from the bulk phase. In general, the grain boundaries (GBs) are known to affect the overall conductivity of multifarious materials either in a positive or negative way. They also assert a considerable impact on the ion transport kinetics and these effects are not yet fully cognized.

Dawson et al. [44] have probed the effect of resistance offered by grain boundaries, thereby hindering the mobility of Li^+ ions in Li_3OCl from DFT and MD studies. They have estimated higher negative boundary energies, which is a direct indication of large concentrations of GBs in Li_3OCl. Moreover, they have estimated the ionic conductivity of the bulk material at 500 K as 6.55×10^{-3} S/cm. Besides, it was observed that the thermodynamically stable GBs exhibited Li-ion conductivities in the order of 10^{-4} S/cm, which is lower than that of the bulk, indicating the prevalence of grain boundary resistance. In addition, a higher migration energy was noticed for the GB (0.56 eV) when compared to that of the bulk (0.29 eV). Meanwhile, Shen et al. [45] proposed a combination of phase field modelling and DFT approach to characterise GBs. It has been found that coherent grains with larger size enhanced the diffusion of Li ions in Li_3OCl. In the case of $Li_2OHCl_{1-x}Br_x$, it has been noticed that the ionic conductivity estimated from Arrhenius plots decreases by two orders of magnitude for the diffusion of Li ions at the GB compared to that of the grains [46]. Thus, from these two studies, it is conspicuous that well-ordered larger grains enhance the ionic conductivity of the ions of interest. Furthermore, Quirk et al. [47] cognized the role of GBs in three Li-based SEs including Li_3InCl_6, Li_3OCl and Li_3PS_4. The evaluation of electronic structure for these three materials revealed reductions in the forbidden energy gap values, which might accelerate dendrite formation. It has been observed that localised trap states present in these materials are responsible for this outcome and better doping strategies can be adopted in order to minimise these states. Moreover, the value of electrostatic potential (ESP) was plotted against the distance from the grain boundary using MD simulations. Interestingly, huge perturbations in ESP were observed for Li_3OCl, and this served as an explanation for the high activation energy demanded by Li^+ ions in this structure. Conversely, flat ESP profiles at GBs of materials including Li_3InCl_6 and Li_3PS_4 explained their minor influence on the conductivity of Li^+ ions.

4.2 Cognizing the solid electrode-solid electrolyte interface from an interfacial resistance viewpoint

Generally, there are space charge layers (SCL) created at the interface to balance the difference in electrochemical potential between the electrode and electrolyte owing to the presence of multiple charge carriers such as electrons and ions. This layer produces a concentration difference with a higher number of ions in one zone and depletion of ions in the other, thereby facilitating the interfacial resistance.

Haruyama et al. [48] carried out a systematic study and modelled cathode/electrolyte ($LiCoO_2/Li_3PS_4$) and cathode/buffer/electrolyte ($LiCoO_2/LiNbO_3/Li_3PS_4$) systems to accurately mimic the solid-solid interface and evaluate the role of SCL. Besides, they have employed DFT + U calculations and estimated the electronic structures in addition to the prediction of Li vacancy formation energetics (E_{vf}). In the case of the cathode/electrolyte interface, it was observed that Li preferred to occupy the ridge sites of CoO_6 octahedra, thereby leading to a distorted interface resembling the SCL. In contrast to this, the introduction of $LiNbO_3$ buffer between the electrode/electrolyte interface suppressed the inhomogeneous adsorption of Li ions. Also, evaluation of Li chemical potential from E_{vf}, indicated the transfer of Li ions under low voltage conditions, which might enhance the interfacial resistance. However, the interposed buffer layer has been found to suppress SCL and offer multiple pathways for the transport of Li^+ ions.

Kim et al. [49] have also demonstrated a meticulous computational approach to characterise the interfacial properties of SEs. In addition, a range of properties at the Li_3OCl/Li interface including wettability, interfacial energy, electrochemical window, contact angle for adhesion and band edge shifts have been estimated. Moreover, the contact angle has been measured from the work of adhesion and surface potential of Li employing the Young-Duprè equation. It had been discovered that Li will wet Li_3OCl electrode when it is inclined at a contact angle of 53°, thereby ensuring strong bonding and lesser interfacial resistance. This bonding at the interface increases the valence band maximum (VBM) and lowers the conduction band minimum (CBM) by bringing down the band gap from 8 to 5.6 eV. Nevertheless, the CBM of Li_3OCl is situated at ~1 eV, denoting its stability against reduction by Li metal anode.

To summarise, the different computational methods explored in the previous sections in order to investigate the properties of electrodes, electrolytes and interfaces, based on the measured parameters, have been consolidated in **Table 1**.

System under study	Computational method	Properties investigated	Key parameters measured
Electrode	Density functional theory (DFT)	Structural stability	Cohesive, formation and Gibbs free energies
	Density functional theory	Reaction voltage	Intercalation/deintercalation voltages
	Cluster expansion (CE) technique with Monte Carlo simulation	Voltage profile and formation energy	
	Density functional theory	Electronic properties	Density of states, band structure charge density difference plots and bader charge analysis
	Integrated Crystal Orbital Hamilton Population (ICOHP) analysis	Bond strength and nature of bonding	
	Electron localisation function (ELF)	Interaction between the atoms	
	Nudged elastic band (NEB) simulation	Electrochemical activity	Activation energy barrier and ionic conductivity
	Molecular dynamics (MD) simulation	Ion transport kinetics	Ionic conductivity and diffusion properties
	Adsorption kinetics	Electrode reactivity	Reaction mechanism
Electrolyte	Density functional theory	Thermodynamic and electrochemical stabilities	Formation energies with and without defects, Gibbs free energies
	Density functional theory	Mechanical stability	Elastic constants
	Ab initio molecular dynamics (AIMD) simulation	Dynamical stability	Phonon dispersion spectra
	Density functional theory	Electronic properties	Band gap and electronic conductivity
	MD simulation	Ion transport properties	Ionic conductivity
Interface	Phase field modelling, DFT and MD approaches	Characterisation of boundary surface	Impact of grain boundary surfaces from diffusion coefficient values and electrostatic potential plot
	DFT + U calculation	Interfacial properties	Space charge layers, wettability, interfacial energy, contact angle for adhesion and band edge shifts

Table 1.
Overview of computational approaches and calculated parameters for investigating the properties of electrodes, electrolytes and interfaces.

5. Challenges and perspectives

Although extensive research has been carried out on the design of electrodes and electrolytes, computational studies focusing on the electrode/electrolyte interface are comparatively limited. This is solely because of the immense model systems, the suitability of the designed models with respect to the ones observed experimentally (especially the structure of grains and GBs is not available for most of the systems), as well as the computational laboriousness associated with them. Apart from these, the interface is one of the most active regions with mobile charge carriers, which again poses a serious challenge. As a conclusion, we furnish the strategies that can be adopted to overcome these issues as follows.

Firstly, modelling the interfacial reactions requires the usage of a larger length scale, which is beyond the scope of AIMD simulations. Thus, modelling this highly interactive region necessitates the employment of hybrid ab initio reactive MD (HAIR-MD) methodology, which combines AIMD with reactive force fields [50]. Since electrons are typically not taken into account in the ReaxFF framework, AIMD simulation can be employed to study electrochemical reactions involving electron transport. Besides, the reactive MD (RMD) technique can be used to tackle the mass transfer processes involving layer formation (interfacial and space charge at the anode and cathode) that demand additional time in the range of nanoseconds. The outcome from these studies includes the reaction energies, bond lengths, bond angles and pathways, which might shed critical insights into the mechanistic interactions behind the interface formation in SSBs.

Secondly, we can estimate certain parameters using DFT, AIMD as well as RMD and apply these values to equations based on electrochemical phenomena or transport as implemented in the Multiphysics-based simulation software, COMSOL. Moreover, the application of boundary conditions with respect to potential and concentration difference can be achieved. Further, executing the simulation will shed significant insights into the interfacial models.

Furthermore, the governing factors for the work of adhesion can be estimated using machine learning (ML) algorithms. Here, the data can be derived from lattice, electronic, mechanical, thermodynamic and chemical perspectives and subjected to feature filtering. This can be done to reduce the model size and eliminate issues with respect to overfitting. Various ML algorithms can be applied and the effectiveness of each model should be assessed. Further, the methods such as the wrapper can be employed to identify the key factors influencing adhesion. Considering the context of SSBs, it is imperative to determine the interfacial properties such as wettability, contact angle and work of adhesion, using machine learning and graph neural networks models. The design and implementation of Hybrid MD and ML-based interfacial models will accelerate the discovery of promising SSB chemistries that are experimentally synthesisable, thereby offering their fair share of contributions in transforming the research challenge of the early 2020s into the battery technology (SSB) of the 2030s.

6. Conclusion

In summary, this chapter examined the computational methodologies used to analyse the electrodes, electrolytes and interfaces of Li as well as post-Li-ion batteries. It also discussed the key parameters identified through various approaches to assess the structural stability, reaction voltage, electronic structure and ionic conductivity of potential battery materials. Additionally, the importance of solid-state electrolytes in batteries, along with their characteristics as studied using DFT and MD methods to evaluate their functionality, was highlighted. Furthermore, the extension of these methods to investigate the interfacial properties and overcome challenges related to grains, grain boundaries, etc., was explored. The chapter also revealed the difficulties in modelling electrode/electrolyte interfaces and suggested the use of the hybrid ab initio reactive MD (HAIR-MD) method to better understand interfacial processes. Overall, the development of hybrid MD and machine learning-based interfacial models will significantly expedite the identification of promising SSB chemistries, thereby paving the way for transformative advancements in battery technology.

Acknowledgements

The authors thank SRM Institute of Science and Technology, Kattankulathur, India, for providing the facilities and resources. Also, we thank the Institute of Electrochemistry-Ulm University, Helmholtz-Institute Ulm Electrochemical Energy Storage and Karlsruhe Institute of Technology, Germany, for their constant support.

Conflict of interest

There are no conflicts of interest to declare.

Author details

Hepsibah Priyadarshini Christopher[1,2,3] and Sudha Venkatachalam[4*]

1 Institute of Electrochemistry, Ulm University, Ulm, Germany

2 Helmholtz-Institute Ulm (HIU) Electrochemical Energy Storage, Ulm, Germany

3 Karlsruhe Institute of Technology (KIT), Karlsruhe, Germany

4 Department of Chemistry, SRM Institute of Science and Technology, Kattankulathur, India

*Address all correspondence to: sudha.venkatachalam@hotmail.com and sudhav1@srmist.edu.in

References

[1] Goodenough JB, Park K. The Li–ion rechargeable battery: A perspective. Journal of the American Chemical Society. 2012;**135**:1167-1176. DOI: 10.1021/ja3091438

[2] Tarascon JM, Armand M. Issues and challenges facing rechargeable lithium batteries. Nature. 2001;**414**:359-367. DOI: 10.1038/35104644

[3] Kulova TL, Fateev VN, Seregin EA, Grigoriev AS. A brief review of post–lithium–ion batteries. International Journal of Electrochemical Science. 2020;**15**:7242-7259. DOI: 10.20964/2020.08.22

[4] FAA Office of Security and Hazardous Materials Safety. Available from: https://www.fire.tc.faa.gov/pdf/TC-TN16-22.pdf

[5] Fire Report: How & Why Do Lithium–Ion Batteries Fail, Insight from the Jedi Master of Lithium Power! 2020. Available from: https://www.waste360.com/waste-recycling/april-2020-fire-report-how-why-do-lithium-ion-batteries-fail-insight-from-the-jedi-master-of-lithium-power-

[6] Schwietert TK, Vasileiadis A, Wagemaker M. First–principles prediction of the electrochemical stability and reaction mechanisms of solid–state electrolytes. JACS Au. 2021;**1**:1488-1496. DOI: 10.1021/jacsau.1c00228

[7] Shinde SS, Wagh NK, Kim SH, Lee JH. Li, Na, K, Mg, Zn, Al, and Ca anode interface chemistries developed by solid–state electrolytes. Advancement of Science. 2023;**10**:e2304235, 1-76. DOI: 10.1002/advs.202304235

[8] He Q, Yu B, Li Z, Zhao Y. Density functional theory for battery materials. Energy and Environmental Materials. 2019;**2**:264-279. DOI: 10.1002/eem2.12056

[9] Ma Y. Computer simulation of cathode materials for lithium ion and lithium batteries: A review. Energy and Environmental Materials. 2018;**1**:148-173. DOI: 10.1002/eem2.12017

[10] Arroyo-deDompablo ME, Dominko R, Gallardo-Amores JM, Dupont L, Mali G, Ehrenberg H, et al. On the energetic stability and electrochemistry of Li_2MnSiO_4 polymorphs. Chemistry of Materials. 2008;**20**:5574-5584. DOI: 10.1021/cm801036k

[11] Zhu L, Zeng YR, Wen J, Li L, Cheng TM. Structural and electrochemical properties of Na_2FeSiO_4 polymorphs for sodium–ion batteries. Electrochimica Acta. 2018;**292**:190-198. DOI: 10.1016/j.electacta.2018.09.170

[12] Mayo M, Griffith KJ, Pickard CJ, Andrew J. Morris Ab initio study of phosphorus anodes for lithium– and sodium–ion batteries. Chemistry of Materials. 2016;**28**(7):2011-2021. DOI: 10.1021/acs.chemmater.5b04208

[13] Zhou F, Kang K, Maxisch T, Ceder G, Morgan D. The electronic structure and band gap of $LiFePO_4$ and $LiMnPO_4$. Solid State Communications. 2004;**132**:181-186. DOI: 10.1016/j.ssc.2004.07.055

[14] Seo DH, Urban A, Ceder G. Calibrating transition metal energy levels and oxygen bands in first principles calculations: Accurate prediction of redox potentials and charge transfer in lithium transition metal oxides. Physical Review B. 2015;**92**:115118, 1-11. DOI: 10.1103/PhysRevB.92.115118

[15] Meng YS, Arroyo-de Dompablo ME. First principles computational materials design for energy storage materials in lithium–ion batteries. Energy & Environmental Science. 2009;**2**:589-609. DOI: 10.1039/B901825E

[16] Arroyo-de Dompablo ME, Krich C, Nava-Avendano J, Palacin MR, Barde F. In quest of cathode materials for Ca ion batteries: The $CaMO_3$ perovskites (M = Mo, Cr, Mn, Fe, Co and Ni). Physical Chemistry Chemical Physics. 2016;**18**:19966-19972. DOI: 10.1039/C6CP03381D

[17] Xiong L, Hu J, Yu S, Wu M, Xu B, Ouyang C. Density functional theory prediction of Mg_3N_2 as a high–performance anode material for Li–ion batteries. Physical Chemistry Chemical Physics. 2019;**21**:7053-7060. DOI: 10.1039/C8CP07398H

[18] Sudarsanan V, Augustine AM, Ravindran P. Investigation of electronic structure and electrochemical properties of Na_2MnSiO_4 as a cathode material for Na ion batteries. Journal of Physical Chemistry C. 2021;**125**(47):25968-25982. DOI: 10.1021/acs.jpcc.1c07386

[19] Dixit M, Markovsky B, Aurbach D, Major DT. Unraveling the effects of Al doping on the electrochemical properties of $LiNi_{0.5}Co_{0.2}Mn_{0.3}O_2$ using first principles. Journal of the Electrochemical Society. 2017;**164**:A6359-A6365. DOI: 10.1149/2.0561701jes

[20] Dua H, Deb J, Paul D, Sarkar U. Twin–graphene as a promising anode material for Na–ion rechargeable batteries. Journal of Energy Storage. 2022;**5**:1-8. DOI: 10.1021/acsanm.1c00460

[21] Zhao X, Zhao Y, Liu Z, Yang Y, Sui J, Wang HE, et al. Synergistic coupling of Lamellar $MoSe_2$ and SnO_2 nanoparticles via chemical bonding at interface for stable and high–power sodium–ion capacitors. Journal of Chemical Engineering. 2018;**354**:1164-1173. DOI: 10.1016/j.cej.2018.08.122

[22] Wu T, Zhu K, Qin C, Huang K. Unravelling the role of structural water in bilayer V_2O_5 during Zn^{2+} intercalation: Insights from DFT calculations. Journal of Materials Chemistry A. 2019;**7**:5612-5620. DOI: 10.1039/C8TA12014E

[23] Bhattacharya J, Van der Ven A. First–principles study of competing mechanisms of non-dilute Li diffusion in spinel $LixTiS_2$. Physical Review B. 2011;**83**:144302, 1-9. DOI: 10.1103/PhysRevB.83.144302

[24] Mueller T, Hautier G, Jain A, Ceder G. Evaluation of Tavorite-structured cathode materials for lithium-ion batteries using high-throughput computing. Chemistry of Materials. 2011;**23**:3854-3862. DOI: 10.1021/cm200753g

[25] Torres A, Luque FJ, Tortajada J, Arroyo-de Dompablo ME. DFT investigation of Ca mobility in reduced–perovskite and oxidized–marokite oxides. Energy Storage. 2019;**21**:354-360. DOI: 10.1016/j.ensm.2019.06.002

[26] Ma Y, Lockwood GK, Garofalini SH. Development of a transferable variable charge potential for the study of energy conversion materials FeF_2 and FeF_3. Journal of Physical Chemistry C. 2011;**115**:24198-24205. DOI: 10.1021/jp207181s

[27] Ma Y, Garofalini SH. Atomistic insights into the conversion reaction in iron fluoride: A dynamically adaptive force field approach. Journal of the American Chemical Society. 2012;**134**:8205-8211. DOI: 10.1021/ja301637c

[28] He K, Lin F, Zhu Y, Yu X, Li J, Lin R, et al. Sodiation kinetics of metal oxide conversion electrodes: A comparative study with lithiation. Nano Letters. 2015;**15**:5755-5763. DOI: 10.1021/acs.nanolett.5b01709

[29] Hepsibah Priyadarshini C, Sudha V, Harinipriya S. Computational mechanistic insights on Ag_2O as a host for Li in lithium–ion battery. Physical Chemistry Chemical Physics. 2022;**24**:16112-16124. DOI: 10.1039/D2CP01674E

[30] Kiran GK, Periyasamy G, Kamath PV. Role of alloying in Cu_2O conversion anode for Li–ion batteries. Theoretical Chemistry Accounts. 2019;**138**:1-10. DOI: 10.1007/s00214-018-2412-z

[31] Hepsibah Priyadarshini C, Sudha V, Harinipriya S. A DFT approach and perspective of sodiation in Ag_2O host–exploration towards sodium batteries. Journal of the Electrochemical Society. 2023;**170**:080525, 1-12. DOI: 10.1149/1945-7111/acf0ec

[32] Singh HK, Zhang Z, Opahle I, Ohmer D, Yao Y, Zhang H. High-throughput screening of magnetic antiperovskites. Chemistry of Materials. 2018;**30**:6983-6991. DOI: 10.1021/acs.chemmater.8b01618

[33] Deng Z, Zhenbin W, Chu IH, Luo J, Ong SP. Elastic properties of alkali superionic conductor electrolytes from first principles calculations. Journal of the Electrochemical Society. 2016;**163**(2):A67-A74. DOI: 10.1149/2.0061602jes

[34] Chuang L, Lin Z, Dong Y. Lattice dynamics of lithium anti–perovskite solid–state electrolytes Li (X = Cl, Br): An insight into thermal management from first–principles study. Journal of Energy Storage. 2024;**101**:113795, 1-10. DOI: 10.1016/j.est.2024.113795

[35] Han F, Westover AS, Yue J, Fan X, Wang F, Chi M, et al. High electronic conductivity as the origin of lithium dendrite formation within solid electrolytes. Nature Energy. 2019;**4**:187-196. DOI: 10.1038/s41560-018-0312-z

[36] Marana NL, Sgroi MF, Maschio L, Ferrari AM, D'Amore M, Casassa S. Computational characterization of β–Li_3PS_4 solid electrolyte: From bulk and surfaces to nanocrystals. Nanomaterials. 2022;**12**(16):2795, 1-13. DOI: 10.3390/nano12162795

[37] Fang H, Jena P. Sodium superionic conductors based on clusters. ACS Applied Materials & Interfaces. 2019;**11**:963-972. DOI: 10.1021/acsami.8b19003

[38] Yu Y, Wang Z, Shao G. Theoretical tuning of Ruddlesden-Popper type anti-perovskite phases as superb ion conductors and cathodes for solid sodium ion batteries. Journal of Materials Chemistry A. 2019;**7**:10483-10493. DOI: 10.1039/C9TA02166C

[39] Fang H, Jena P. Li-rich antiperovskite superionic conductors based on cluster ions. Proceedings. National Academy of Sciences. United States of America. 2017;**114**(42):11046-11051. DOI: 10.1073/pnas.1704086114

[40] Tsai PC, Mair S, Smith J, Halat DM, Chien PH, Kim K, et al. Double paddle-wheel enhanced sodium ion conduction in an antiperovskite solid electrolyte. Advanced Energy Materials. 2022;**13**:2203284. DOI: 10.1002/aenm.202203284

[41] Lü X, Wu G, Howard JW, Chen A, Zhao Y, Daemen LL, et al. Li-rich anti-perovskite Li_3OCl films with

enhanced ionic conductivity. Chemical Communications. 2014;**50**(78):11520-11522. DOI: 10.1039/C4CC05372A

[42] Kim K, Siegel DJ. Multivalent ion transport in anti-perovskite solid electrolytes. Chemistry of Materials. 2021;**33**:2187-2197. DOI: 10.1021/acs.chemmater.1c00096

[43] Chao Z, Bingzheng S, Zhaohui Q, Xuejie L, Yuan R. The effect of defects for the ion transport of Li_3ScCl_6 and Li_3InCl_6 with the Interface of lithium metal anode: A first-principles study. Materials Today Communications. 2023;**35**:105764, 1-13. DOI: 10.1016/j.mtcomm.2023.105764

[44] Dawson JA, Canepa P, Famprikis T, Masquelier C, Islam MS. Atomic-scale influence of grain boundaries on Li-ion conduction in solid electrolytes for all-solid-state batteries. Journal of the American Chemical Society. 2018;**140**:362-368. DOI: 10.1021/jacs.7b10593

[45] Shen K, Wang Y, Zhang J, Zong Y, Li G, Zhao C, et al. Revealing the effect of grain boundary segregation on Li ion transport in polycrystalline anti-perovskite Li_3ClO: A phase field study. Physical Chemistry Chemical Physics. 2020;**22**:3030-3036. DOI: 10.1039/C9CP06055C

[46] Lee HJ, Darminto B, Narayanan S, Diaz-Lopez M, Xiao AW, Chart Y, et al. Li-ion conductivity in $Li_2OHCl_{1-x}B_x$ solid electrolytes: Grains, grain boundaries and interfaces. Journal of Materials Chemistry A. 2022;**10**:11574-11586. DOI: 10.1039/D2TA01462A

[47] Quirk JA, Dawson JA. Design principles for grain boundaries in solid-state lithium-ion conductors. Advanced Energy Materials. 2023;**13**:2301114. DOI: 10.1002/aenm.202301114

[48] Haruyama J, Sodeyama K, Han L, Kazunori T, Yoshitaka T. Space-charge layer effect at interface between oxide cathode and sulfide electrolyte in all-solid-state lithium-ion battery. Chemistry of Materials. 2014;**26**(14):4248-4255. DOI: 10.1021/cm5016959

[49] Kim K, Siegel DJ. Predicting wettability and the electrochemical window of lithium-metal/solid electrolyte interfaces. ACS Applied Materials & Interfaces. 2019;**11**:39940-39950. DOI: 10.1021/acsami.9b13311

[50] Liu Y, Yu P, Wu Y, Yang H, Xie M, Huai L, et al. The DFT-ReaxFF hybrid reactive dynamics method with application to the reductive decomposition reaction of the TFSI and DOL electrolyte at a lithium-metal anode surface. Journal of Physical Chemistry Letters. 2021;**12**(4):1300-1306. DOI: 10.1021/acs.jpclett.0c03720

Chapter 7

Novel Technologies for the Next-Generation of Vehicle Battery Systems

Karsten Mueller, Michael Clauss, Alexander Fandakov, Johannes Werfel and Oliver Nolte

Abstract

Despite significant advancements in the development of battery systems, which have enhanced volume utilization and vehicle range, sales in some regions are currently stagnating due to high prices and range anxiety. Over the past decade, the authors have been involved in the development of a compact bipolar Li-ion battery system, engineered to achieve driving ranges of up to 1000 km by integrating the battery with its alternative cell setup directly into the vehicle chassis. The planar bipolar battery features foils with anodes and cathodes on opposite sides (therefore called bipolar), connected in series to create a flat, space-efficient design. By eliminating the need for traditional module organization, this technology aims for high energy densities, low costs, and minimized internal resistance for enhanced efficiency. However, the bipolar concept also introduces unique technical challenges, such as the need for specialized electrical design and dedicated manufacturing equipment. This contribution explores the latest advancements in the authors' work on bipolar battery development, additionally emphasizing the potential of implementing solid-state electrolytes, highlighting the opportunities for innovative platform concepts in vehicle applications. By optimizing battery design and materials, the development of bipolar lithium-ion batteries holds the promise of revolutionizing energy storage solutions, paving the way for more efficient, cost-effective, and sustainable electric vehicle technology.

Keywords: bipolar battery, lithium-ion, solid-state, energy density, novel battery technologies

1. Introduction

In recent years, significant investments have been made in the development of specialized platforms for battery electric vehicles (BEVs) to maximize volume utilization and enhance vehicle aerodynamics. Despite these advancements, the sales of BEVs have stagnated in several key markets, failing to meet the ambitious targets set by governments worldwide. Surveys and analyses suggest that this stagnation is primarily due to persistently high prices and, in some cases, range anxiety.

Concurrently, the tightening of CO_2 fleet limits is exerting additional pressure on the automotive industry to innovate and adapt [1, 2].

In response to these challenges, IAV initiated the development of a bipolar lithium-ion battery system named "EMBATT" Chassis-embedded Energy in 2014. This project represents a significant improvement in the state-of-the-art battery systems for BEVs. The primary objective of this concept is to create more efficient, next-generation battery systems capable of achieving electric ranges of up to 1000 kilometers [3]. It emphasizes integration at the system-to-cell level and the elimination of interfaces. It is accomplished by maximizing the proportion of active material within the entire battery volume and minimizing the space required for passive components. Therefore, it agrees with the general trend of the market (e.g., BYD Blade prismatic cells [4]) but utilizes a different cell configuration to achieve even higher levels of volume utilization. This alternative approach is accompanied by partially unique technical challenges compared to conventional cell, module, and system architecture, respectively.

This contribution initially provides an overview of the concept of bipolar batteries and their concept-specific challenges, then focuses on cell-specific challenges and the anticipated potential. It addresses the selection of suitable materials for use as bipolar foil, which must electrochemically withstand a wide potential range as the anode as well as the cathode material is coated on one side, respectively. Additionally, the use of solid electrolytes for the fabrication of the separator layer and the production of electrodes is examined. Finally, a developed prototype production line for bipolar battery stacks is presented.

2. The concept of bipolar batteries

Traditional battery systems, which rely on the standard cell configuration of active material layers connected in parallel and encapsulated within a cylindrical, prismatic can, or pouch bag, often necessitate complex module and pack organization to achieve the requirements on system level (energy, voltage, power). This leads to substantial unused space. In contrast, the bipolar approach eliminates these requirements, by directly connecting active material layers in a serial stack to achieve the target voltage. This allows a fundamentally different structural design approach aimed at maximizing volumetric efficiency at the system level due to elimination of traditional cell housing and module organization. Additionally, it enables the implementation of chassis-embedded technology, integrating the vehicle chassis as battery housing.

2.1 Motivation and working principle

The planar bipolar battery technology targets addressing major pain points of state-of-the-art battery systems. First, the technology aims to achieve a volumetric energy density of 500 Wh/l at the system level, significantly enhancing the values typical for conventional systems [5, 6]. This is accomplished by maximizing the fraction of active material in the overall system volume while reducing the space required for passive components (**Figure 1**). The maximized utilization of available volume is enabled by combining increased energy density at both the cell (bipolar configuration and maximized single cell size) and system (chassis-embedded approach with platform and energy storage system integrated into a single unit) levels.

In addition, it seeks to reduce overall system costs to approximately 100 €/kWh, making the technology more cost-effective and accessible, thus suitable for the mass

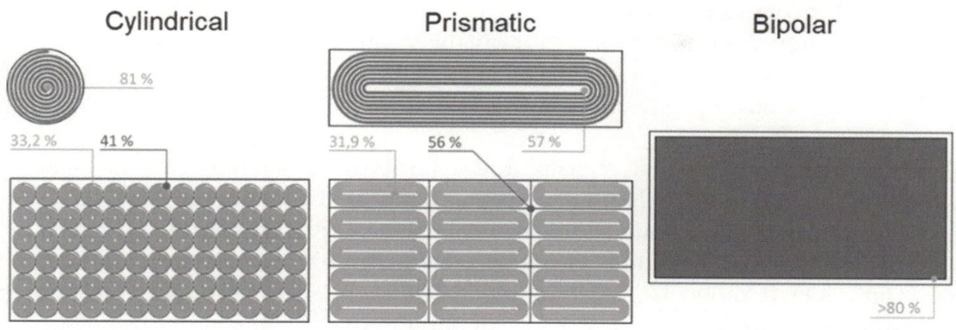

Figure 1.
Typical volume fractions utilized for storing electrical energy in common and bipolar cell formats [7].

market. Furthermore, bipolar technology enables a lower internal battery resistance of up to one-third compared to state-of-the-art battery systems, thus enabling significantly improved performance, efficiency, and ultimately vehicle range.

The core component of the planar bipolar battery, as illustrated in **Figure 2** right, is the bipolar foil with its active material coatings. These foils are assembled into a stack, with each foil featuring an anode on one side and a cathode on the other. All electrochemical cells are isolated from each other with seals to prevent ion flux. Electrical connections to the outside are made through two metal endplates. This structural organization allows the cells to be connected in series, achieving a high voltage level already on cell level while having a very flat design. The traction system voltage is determined by the unit cells voltage multiplied by the number of cells in the stack.

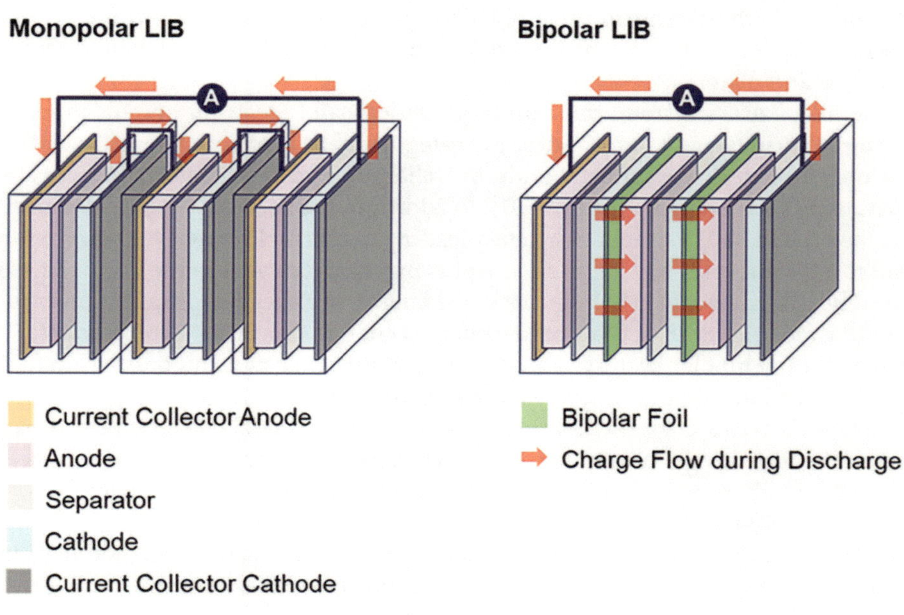

Figure 2.
Mono- (left) and bipolar (right) principle overview [7].

Thanks to this compact design and the elimination of interfaces, such as cell connecting bus bars, superior utilization of the available volume can be achieved.

2.2 Overview of concept-specific challenges

The specifics of the bipolar concept described above present unique challenges, with development heavily relying on reliably capturing the interaction between electrical, chemical, thermal, and mechanical domains. Battery cells are typically equipped with coated metal foils as voltage tabs, with aluminum used for the cathode and copper for the anode. However, bipolar battery cells cannot use solely one of these materials as their foils due to the reaction of aluminum with lithium at low potentials vs. Li/Li^+ and copper dissolution at high potentials vs. Li/Li^+, necessitating the use of a material with a high electrochemical stability for both active materials potential ranges [8]. Conductive polymer foil is an alternative solution since it works with typical anode and cathode potentials. A major disadvantage of polymer foil, however, is its low electrical conductivity. The goal is hence to reduce the voltage drop of the polymer foil, optimizing the current flow within the cell and minimizing losses. A polymer foil with the necessary electrical conductivity has not yet been identified. Alternatively, metal-based multimaterial solutions can be employed as a compromise.

In a parallel connected unit, cells in a stack show an implicit effect of balancing aging between the unit cells to a certain extent. If the resistance of one cell is increased due to aging (e.g., solid electrolyte interface formation), other cells experience higher local current densities due to Ohm's law, which leads to higher aging in these cells. In a bipolar battery cell, this mechanism can only occur within a unit cell with local inhomogeneities, but not between cells. Thus, if one cell ages faster compared to other cells (e.g., due to worse cooling conditions, manufacturing fluctuations, etc.), a higher impact on the whole cell stack can be observed [9]. Therefore, it is crucial to achieve as homogeneous cell behavior as possible.

The stacked architecture of the bipolar battery and its generally high surface area necessitate a more-thorough consideration of the electrical fields and conductivity within foils and plates.

Generally, battery systems must undergo critical safety tests for homologation, adhering to various regulations. These are categorized as mechanical (e.g., crush, vibration), electrical (e.g., short circuit, overcharge), and thermal tests (e.g., overtemperature, thermal propagation) [10]. In addition, the thermal propagation test prescribed in GB 38031 simulates a defect leading to thermal runaway, observing the spread to adjacent cells, aiming to ensure passenger safety event in the occurrence of this highly critical event [11]. As the released energy in a thermal propagation case can be 10 times higher than the stored electrical energy, this poses a considerable challenge for the bipolar battery with its highly compact design and only a thin foil separating a cell in thermal runaway from neighboring cells [12].

The bipolar technology further necessitates the use of specialized manufacturing equipment and redesigned coating processes to accommodate the specific requirements, with facilitating scalability in production considered a critical factor for commercial viability.

Due to high material, manufacturing, and testing costs, high-fidelity simulation approaches are essential to accurately model and optimize these batteries in the design process to tackle the mentioned challenges.

Ultimately, employing a bipolar battery for vehicle applications requires and enables the development of completely new platform concepts.

3. Cell-specific challenges and potentials

3.1 Bipolar foils

In the field of lithium-ion battery technology, the choice of conductor materials is important to ensure optimal performance, durability, and efficiency. These materials must meet stringent requirements, including chemical resistance to electrolytes and salts, high electrical conductivity, and electrochemical stability across a broad potential range [13]. Various materials, compositions, and process variants are being studied to optimize the performance of conductor materials [7].

Bipolar foils, which are in contact with the active materials of the anode and cathode, must be inert to side reactions within the whole potential range of both electrodes to ensure the longevity of bipolar batteries. Although aluminum is commonly used, it is prone to corrosion through alloy formation at low potential vs. Li/Li^+, which are characteristic at the anode [14]. In 1990, Besenhardt et al. discovered that the addition of copper can significantly enhance the stability of aluminum conductors [15]. Today, copper foils are employed as conductor material of the graphite anode in conventional lithium-ion batteries [16]. However, due to the higher cost of copper compared to aluminum and its instability at high potentials vs. Li/Li^+ (e.g., cathode potentials), it is important to explore alternative materials. In bipolar cell design, the coating of aluminum with thin, closed copper layers is being investigated as a potential solution [17].

Different treatment methods, such as galvanic coating, inkjet coating, screen printing, and metal deposition, are being tested to improve adhesion, layer thickness, and conductivity. Coated aluminum-copper foils have shown electrical stability and flexibility comparable to pure copper, making them a viable alternative. Another promising approach is the use of cost-effective steel foils as carriers for cathode and anode materials, with the requirement that the coating be completely closed to prevent corrosion [7].

Polymer foils, such as polyvinylidene fluoride (PVDF), offer advantages in terms of weight and cost savings and decrease the overall metal consumption. However, untreated polymer foils exhibit poor adhesion and conductivity. Conductivity can be enhanced through nickel coating. Initially, chemical and mechanical roughening of the surface is required to improve surface adhesion (**Figure 3**). Subsequent hydroxylation with LiOH not only enhances surface adhesion but also improves wettability. The improvement in wettability due to chemical modification was experimentally determined through contact angle measurements before and after hydroxylation.

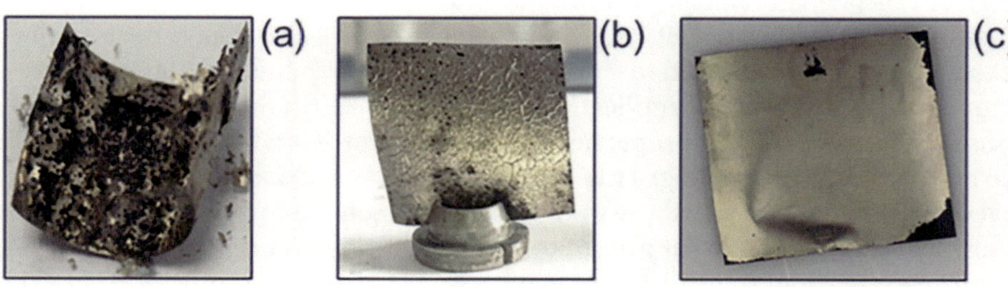

Figure 3.
PVDF processing: (a) Chemical roughening with metallic sodium or sodium salts; (b) mechanical roughening with SiC abrasive fleece; and (c) hydroxylation (LiOH) followed by palladium-catalyzed nickel deposition [7].

The contact angle decreased from 73.9 to 60.5°. A subsequent adhesion test using adhesive tape showed no detachment of the coating, indicating its stability. The coating of the PVDF film with an approximately 130 nm thick Ni layer exhibited a 16-fold increase in specific conductivity compared to the uncoated PVDF film [7].

Electrochemical stability was assessed using cyclic voltammetry to identify potential side reactions within the relevant potential window for the different materials (Ni-coated PVDF, Ni-coated steel, plated aluminum-copper). The plated aluminum-copper variant demonstrated superior performance by offering advantages in complexity, availability, thickness, and costs, leading to its selection as the conductor material for current bipolar cells [5].

3.2 Sense terminals

In the design and operation of lithium-ion battery modules and packs, the implementation of measurement contacts on each cell is crucial for monitoring individual cell voltages and ensuring proper balancing across the system. This monitoring is essential to maintain the health and efficiency of the battery, preventing issues such as overcharging or deep discharging, which can lead to reduced performance or even damage. To achieve this for the bipolar battery concept, various materials, compositions, and processing methods are being explored to optimize the functionality and reliability of these measurement contacts.

Copper cables are commonly used due to their excellent electrical conductivity, which is vital for accurate voltage measurement and low loss balancing. However, they present challenges in terms of stack integrity, as the diameter of the copper conductors can negatively impact the tightness of the stack, and electrochemical stability as elaborated above. This can lead to potential issues with sealing and mechanical stability, necessitating careful consideration in their application.

Alternative materials, such as nickel foils and copper-coated foils made from polyethylene terephthalate (PET) or polyethylene naphthalate (PEN), offer promising solutions. These materials exhibit good processability and adhesion to sealing tapes, making them suitable for integration into battery stacks. The use of nickel-coated steel foil is another viable option, providing good availability and processability without the need for additional roughening. This material also ensures effective sealing with adhesive tape, contributing to the overall integrity and performance of the battery stack [7].

Several processing methods have been investigated to optimize the application of measurement contacts. Welding, while effective in creating strong connections, presents limitations due to the possible melting of conductors, especially in the case of polymer foils, which can compromise the structural integrity of the stack. As a result, alternative methods, such as adhesives and adhesive tapes, have been explored. In test series, adhesives have led to rippling of the polymeric foil, highlighting the need for careful selection and application. Adhesive tapes, on the other hand, have demonstrated good adhesion properties and low resistance, making them a preferred choice for maintaining the integrity of the battery stack. Final improvement was achieved through the combined use of tape and folding techniques. Folding the sense taps through a cut in the bipolar foil led to minimized contact resistance (**Figure 4**). Based on an evaluation of the different materials in terms of their processability and sealing properties, nickel-plated steel foils with 25 μm and 67 thickness appeared as the preferred option for the battery management system (BMS) connection and were utilized for prototype manufacturing, see Section 3.4 [7].

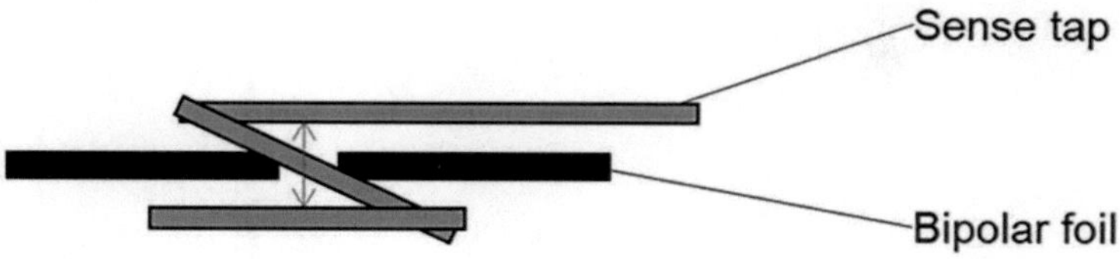

Figure 4.
Folding of sense tap through a cut in bipolar foil [7].

3.3 Solid-state electrolytes

The integration of solid-state electrolytes into the bipolar battery concept offers a promising pathway for further significant enhancements of energy density, thereby enabling even more space- and weight-efficient battery systems. But the prerequisite for their application is the identification of their potential to use them in the production of bipolar battery cells, which was evaluated through various experiments. In the first step, the focus was on the manual fabrication of a separator layer and the production of electrodes. Therefore, the application of a separator layer on aluminum foil and the infiltration of commercial porous electrodes using a solid-state electrolyte solution were investigated [18].

For the application of a separator layer, attempts were made to apply dissolved solid-state electrolytes onto aluminum foil. Polyvinylidene fluoride (PVDF), dissolved in N-methyl-2-pyrrolidone (NMP), was used. Both adhering and detached polymer layers on the aluminum foil could be produced. In this regard, different drying methods influenced the crystallization degree of the formed layers. Higher temperatures lead to faster solvent evaporation and thus less time for the alignment of polymer coils, increasing the proportion of the amorphous phase. Higher proportions of the amorphous phase are desired, as their ionic conductivity is higher than that of crystalline phases. Ideal concentrations for the application of the layers were determined to be in the range of 7 to 10 wt% (PVDF in NMP). The corresponding manufactured layers exhibited a viscosity that allowed easy processing. However, the deformation of the aluminum foil and the formation of defects due to drying were problematic (**Figure 5**). This leads to the requirement of further optimizing the drying process and the weight percent (wt%) of PVDF [18].

In the infiltration experiments, PVDF dissolved in NMP was also used as a solid-state electrolyte. $Li_4Ti_5O_{12}$ (LTO) and graphite electrodes were successfully infiltrated using a PVDF solution (3 wt% in NMP). The verification of infiltration was conducted by qualitative analysis using a microscope (**Figure 6**). Future evaluations are necessary to establish quantitative verification methods, particularly to investigate the influence of vacuum and the concentration of the PVDF solution more precisely.

Furthermore, the use of solid electrolytes enables the application of various anode materials. When using polyethylene glycol (PEG), as solid electrolyte, and NMC811, as the cathode material, different anode materials (graphite, graphite-silicon, lithium metal) were investigated regarding their impact on the cells possible energy density. Lithium metal, with its comparatively high specific capacity of 3860 mAh/g, emerged as the most promising candidate, enhancing the achievable volumetric energy density by approximately 100% compared to the typical liquid electrolyte system (graphite/NMC811) when the same stack size was used [18]. Generally, all further advances in the field of solid-state cells can also be transferred to the bipolar concept as long as a suitable bipolar foil for specific solid-state solutions can be found.

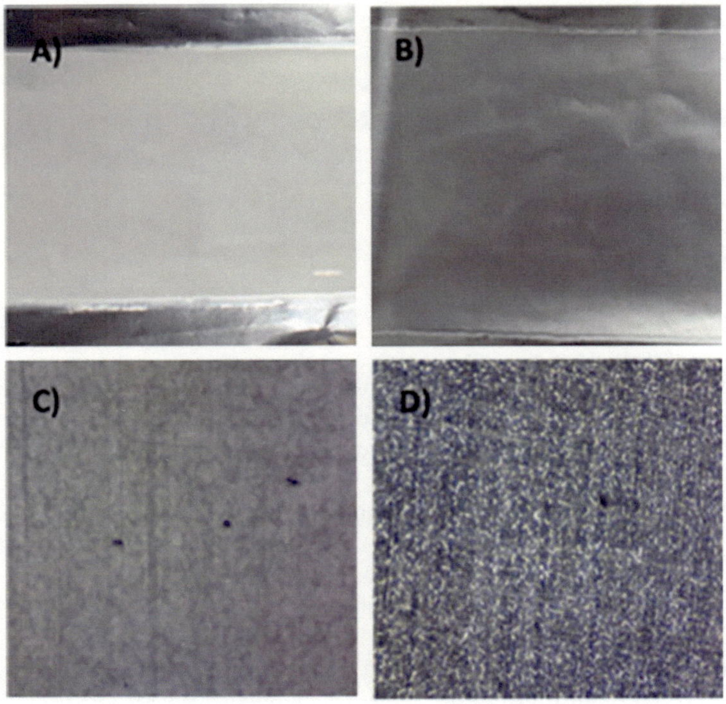

Figure 5.
7 wt% PVDF in NMP: (A) dried in laboratory fume cupboard; (B) dried in oven; (C) microscope picture of (A) shows defects of polymer layer; and (D) microscope picture of (B) shows defects of polymer layer [18].

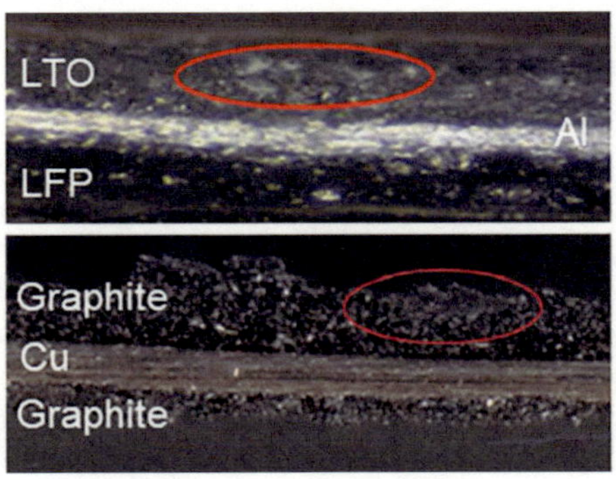

Figure 6.
Microscope micrographs of the infiltration of the LTO anode under atmospheric pressure (top) and of the graphite anode under vacuum at 100 mbar [18].

3.4 Prototype production line

In the pursuit of advancing battery technology, considerable effort has been put into establishing a prototype production line specifically designed for bipolar stacks (**Figure 7**). This production line, capable of handling stacks with dimensions

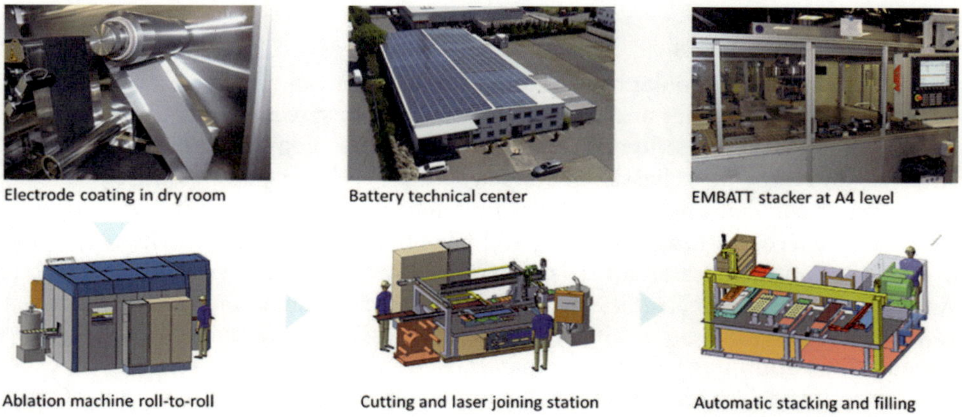

Figure 7.
Prototype production line for bipolar cell stack in the EMBATT research project [7].

of 300 x 1000 mm^2, is tailored to a production process designed to ensure precision and efficiency, addressing the unique challenges associated with bipolar concept. The developed stack manufacturing process has been continuously improved based on insights from evaluations of various manufacturing technologies [3, 7].

The production process includes electrode production, de-coating of active material, laser cutting and joining, stacking, sealing, and electrolyte filling and begins with the application of electrode material onto foils. These are then subjected to a drying process employing a combination of pre-drying at 110°C for 24 hours and vacuum drying at 110°C for 24 hours, respectively, allowing for the residual moisture to be cut in half compared to vacuum drying at 90°C. The drying process must be carefully controlled to prevent any inconsistencies that could affect the electrochemical properties of the electrodes. This step is critical to ensure the uniform distribution and adherence of the electrode material, which directly impacts the performance and reliability of the final battery stack.

Ablation is employed as the next necessary process step to prepare the foil edges for stack sealing. This technique involves the precise removal of material from the edges, creating a clean and defined boundary that facilitates effective sealing, and is crucial for maintaining the integrity of the stack by preventing any leakage or contamination that could compromise the battery's performance. The foils are then separated into individual sheets, and laser technology is used to attach the sense terminals, ensuring these are accurately positioned. Finally, at a third process station, the individual layers are automatically assembled into a complete bipolar stack. Automation in this stage enhances the consistency and speed of production. The filling process involves the introduction of necessary components and materials (e.g., electrolyte, sealings, endplates) to complete the stack, preparing it for integration into larger battery assemblies [7].

Demonstrating the functionality of a prototype production line represents a crucial prerequisite for manufacturing bipolar battery systems by ensuring consistency of the manufactured battery stacks and crucially, facilitating scalability in production, considered a critical factor for commercial viability.

4. Conclusions

The exploration of bipolar battery technology presents a transformative approach to enhancing the efficiency and performance of battery systems. By introducing a fundamentally different structural design, bipolar technology maximizes volumetric efficiency and integrates chassis-embedded solutions, thereby addressing significant limitations of conventional systems. The targeted energy density of 500 Wh/l and reduced system costs to approximately 100 €/kWh underscore the potential for widespread adoption and accessibility in the mass market. The innovative design facilitates lower internal resistance, significantly improving performance and vehicle range. However, the development of bipolar batteries necessitates overcoming challenges, particularly in optimizing conductor materials to enhance corrosion resistance, conductivity, and cost-effectiveness. The implementation of measurement contacts and the exploration of solid-state electrolytes further contribute to the reliability and efficiency of these systems. The establishment of a prototype production line tailored for bipolar stacks demonstrates the feasibility of scaling production, a critical factor for commercial viability. Ultimately, the advancements in bipolar battery technology pave the way for new platform concepts in vehicle applications.

5. Outlook

To show the possible potential of this concept, a passenger car battery system concept was designed, with a capacity of 115 kWh and operating on a 400 V configuration [5, 6]. It integrates the vehicle platform and battery into a single unit through chassis-embedded technology, optimizing spatial utilization and functional integration (**Figure 8**). The battery architecture comprises 112 series and 9 parallel cells (112s9p), housed within a compact format measuring 1000 x 1900 x 115 mm^3. It includes three modules, each containing three stacks, resulting in a total of nine stacks with dimensions of 950 x 590 x 34 mm^3 and a capacity of 31 Ah per stack. A thermal management system is employed, featuring four cooling plates per module to ensure uniform temperature distribution across the stacks. The electrodes utilize graphite anodes and NMC811 cathodes, achieving a volumetric energy density > 450 Wh/l, facilitated by a volume utilization rate of over 75% at Begin of Life (BoL).

The ecological footprint of this battery system is estimated at 66 kg-CO_2-eq. per kWh, underscoring a commitment to environmental sustainability. Economically, the system is cost-effective, with total costs estimated at 105 €/kWh, predominantly

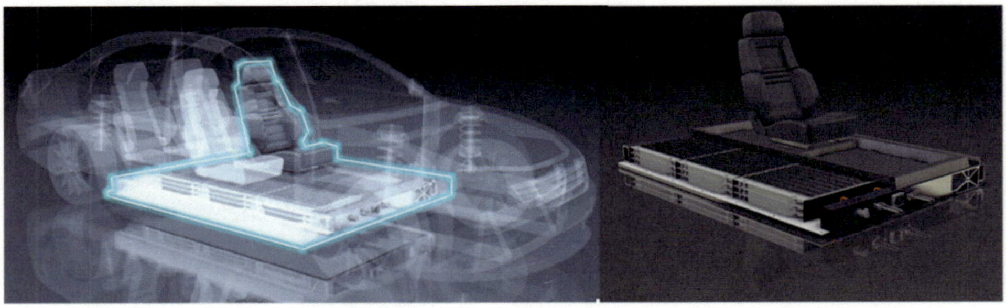

Figure 8.
Final chassis-embedded bipolar platform concept with passenger car battery integration [7].

driven by material costs, particularly the cathode. The battery management system (BMS) is strategically positioned to enable individual cell balancing via a flexible printed circuit board, with a current-carrying capacity designed at 100 mA to ensure effective passive balancing. The use of aluminum and copper conductors further guarantees robust current-carrying capacity.

Future work should focus on further optimizing the materials used in bipolar batteries, particularly in enhancing the electrochemical stability and conductivity of conductor materials. The exploration of alternative materials and advanced manufacturing techniques will be crucial in overcoming current limitations and achieving cost-effective production. Additionally, the development of high-fidelity simulation models will be important for accurately predicting the performance and reliability of bipolar batteries in real-world applications. Furthermore, the behavior of this technology regarding safety aspects (e.g., thermal runaway, crash/crush) should be investigated more thoroughly based on prototypes of a relevant size.

Acknowledgements

The EMBATT concept has been developed in a research project in collaboration with Thyssenkrupp System Engineering and Fraunhofer IKTS. The funding has been provided by the European Regional Development Fund of Saxony, the German Federal Ministry of Education and Research, and the German Federal Ministry for Economic Affairs and Energy.

Conflict of interest

The authors declare no conflict of interest.

Other declarations

The authors disclose the use of AI tools (proprietary implementation of ChatGPT 4.0) as basic author support for refining and correcting parts of this manuscript.

Author details

Karsten Mueller, Michael Clauss, Alexander Fandakov*, Johannes Werfel and Oliver Nolte
IAV GmbH, Berlin, Germany

*Address all correspondence to: alexander.fandakov@iav.de

References

[1] Katanich D. EU Car Industry Calls for Urgent Action as EV New Car Sales Tumble. Euro News; 2024. Available from: https://www.euronews.com/business/2024/09/19/eu-car-industry-calls-for-urgent-action-as-ev-new-car-sales-tumble [Accessed: May 31, 2025]

[2] European Parliamant, Council. Regulation (EU) 2023/851 Amending Regulation (EU) 2019/631 as Regards Strengthening the CO2 Emission Performance Standards for New Passenger Cars and New Light Commercial Vehicles in Line with the Union's Increased Climate Ambition. Brussels. 19 April 2023. Document 32023R0851

[3] EMBATT. Chassis Embedded Energy. Available from: https://embatt.de/ [Accessed: October 31, 2024]

[4] Gorsch J et al. Contrasting a BYD blade prismatic cell and tesla 4680 cylindrical cell with a teardown analysis of design and performance. Cell Reports Physical Science. 2025;**6**(3):102453

[5] Müller K et al. EMBATT – A bipolar battery approach to meet future requirements for extended ranges. In: CTI Europe. Berlin: Car Training Institute - A Division of Euroforum Deutschland GmbH; 2019

[6] Müller K, Clauß M, et al. EMBATT bipolar battery technology. In: Battery Show Europe. Stuttgart: Informa Markets; 2019

[7] Laubenstein R. Next-gen batteries: Bipolar technology for better KPIs at lower costs. In: 42nd International Battery Seminar. Needham: Cambridge EnerTech; 2025

[8] Langner T et al. Studies on the deposition of copper in lithium-ion batteries during the deep discharge process. Scientific Reports. 2021;**11**:6316

[9] Pohle B et al. Design optimization of battery systems by multiphysics simulation. In: International Congress: SIA Power Train & Electronics; 12 & 13 June 2019; Paris. 2019

[10] United Nations. Regulation no. 100. In: Uniform Provisions Concerning the Approval of Vehicles with Regard to Specific Requirements for the Electric Power Train; E/ECE/Rev.2/Add.99/Rev.3. New York City: United Nations; 2022

[11] National Standard of the People's Republic of China. Electric Vehicles Traction Battery Safety Requirements; GB 38031-2020. Peoples Republic of China, Peking; 2021

[12] Sens M et al. From thermal runaway to No thermal propagation. In: 45th International Vienna Motor Symposium. Vienna: Austrian Society of Automotive Engineers (ÖVK); 2024

[13] Myung S-T, Hitoshi Y, Sun Y-K. Electrochemical behavior and passivation of current collectors in lithium-ion batteries. Journal of Materials Chemistry. 2011;**21**:9891-9911

[14] Oltean G, Tai C-W, Edström K, Nyholm L. On the origin of the capacity fading for aluminium negative electrodes in Li-ion batteries. Journal of Power Sources. 2014;**269**:266-273

[15] Besenhardt JO, Hess M, Komenda P. Dimensionally stable Li-alloy electrodes for secondary batteries. Solid State Ionics. 1990;**40-41**:525-529

[16] Korthauer R. Handbuch Lithium-Ionen-Batterien. Heidelberg: Springer Verlag Berlin; 2013. 23-29 p

[17] Jungkurth I. Dünne schichten zum korrosionsschutz von metallischen, elektrischen ableitern in lithium-ionen-batterien [master's thesis]. Chemnitz: TU Chemnitz; 2017

[18] Krüger D. Erstellung eines materialkonzeptes zur integration von festkörperelektrolyten in einer bipolarbatterie [bachelor's thesis]. Nürnberg: Technische Hochschule Nürnberg Georg Simon Ohm; 2019

Chapter 8

Template Methods for the Synthesis of Bimetallic Oxides Used in Lithium-Ion Batteries

Wenming Liao

Abstract

Bimetallic oxides are considered as the alternatives of commercial carbon graphite anode materials of lithium-ion batteries (LIBs) due to their high specific capacity and environmental friendless. Double metal ions have different Li-insertion potentials to buffer volume expansion of each other, thus forming a good synergistic effect for improving their electrochemical performance. However, there are some obstacles to their practical progress in LIBs, such as poor cycle stability and lower delivered capacity. It is necessary to construct porous and uniform bimetallic oxides to promote Li ions intercalation and de-intercalation capability. Porous structure can effectively buffer Li-insertion mechanical deformation, increase electrolyte infiltration, and shorten Li ions diffusion path. Uniform nanoparticles are conducive to smooth reversible Li-insertion transformation reaction and reversible reaction of solid electrolyte interface (SEI) film on the surface of electrode, so as to avoid the overcharge and overdischarge phenomenon caused by uneven particle size. However, the controlled synthesis of bimetallic oxides is more difficult than that of single metal oxides. Template methods are considered as good choices to synthesize porous and uniform bimetallic oxides. Herein, we briefly introduced classic template methods and metal ion-reactive template methods used in the synthesis of nanostructured bimetallic oxides as anodes of LIBs in this chapter.

Keywords: template methods, bimetallic oxides, nanostructured, lithium-ion batteries, electrochemical performance

1. Introduction

Rechargeable lithium-ion batteries (LIBs), as state-of-the-art devices, have been exploited for alleviating one-off energy crisis to satisfy the social development's needs due to their advantages of low cost, environmental friendliness, low self-discharge, high energy density, and no memory, which are used in many electrochemical devices, such as portable electronic equipment, hybrid electric vehicles, and electric vehicles [1–7]. In the running batteries, lithium ions can travel as shuttle ions between the cathodes and anodes [8, 9]. Taken commercial $C/LiPF_6/LiCoO_2$ full batteries as example, the reaction mechanisms of LIBs are listed in the following eqs. (1)–(3). However, traditional graphite anode materials have a low theoretical specific capacity

(372 mAh g^{-1}), thus limiting the development of batteries towards high energy density. Therefore, great efforts have been made to develop new Li-storage materials as the alternatives of commercial graphite carbon materials for used in lithium-ion batteries [10–13].

$$LiCoO_2 \xleftrightarrow{Charge/Discharge} Li_1 - xCoO_2 + xLi^+ + xe \quad (1)$$

$$C_n + xLi^+ + xe \xleftrightarrow{Charge/Discharge} Li_xC_n \quad (2)$$

$$LiCoO_2 + C_n \xleftrightarrow{Charge/Discharge} Li_{1-x}CoO_2 + Li_xC_n \quad (3)$$

Bimetallic oxides are considered as potential anode materials of LIBs due to their lower price, high specific capacity, environmental benignity, and enhanced safety performance [14–18]. Double metal ions have different Li-insertion potentials to buffer the volume expansion of each other, thus forming a good synergistic effect for improving their Li-insertion electrochemical performance. However, there are still some obstacles to their practical progress in LIBs, such as poor cycle stability and lower delivered capacity [14–18]. The main reasons include: (1) the inevitable poor electron/lithium-ion conductivity, (2) the irreversible Li-insertion reaction caused by structural pulverization and collapse, and (3) the unstable solid electrolyte interface (SEI) film and side reactions and excessive consumption of the lithium source in the electrolyte. It is necessary to construct porous and uniform bimetallic oxides to promote Li ions intercalation and de-intercalation capability. Porous structure can effectively buffer the Li-insertion mechanical deformation, increase the adsorption of electrolyte, and shorten the Li ions diffusion path, thereby accelerating Li ions transfer to overcome the major kinetic limitations [14–16]. Uniform nanoparticles are conducive to the smooth reversible transformation reaction and reversible reaction of SEI film on the surface of the material, so as to avoid the overcharge and overdischarge phenomenon caused by uneven particle size, thus improving their electrochemical performance and structural stability [19]. However, in a liquid phase reaction, the interaction of double metal ion valence bond and functional groups in the two precursors affects reaction stability, thereby resulting in a poor controllability in morphological structure of the materials.

Nowadays, classic template methods are regarded as good choices to synthesize porous and uniform bimetallic oxides by using various types of templates as frameworks. They can readily control the morphological structures and sizes of porous and uniform bimetallic oxides, so as to achieve effective control of materials' electrochemical Li-storage properties [20–34]. These templates include polymers, micelles, bubbles, organic compounds, and so on. There are three steps in the classic template methods, namely, (i) the synthesis of templates, (ii) using templates to shape materials, and (iii) templates' removal. In the past more than ten years, many efforts have been devoted to develop a novel metal ion-reactive template method to synthesize the nanostructured bimetallic oxides. The readily synthesized metal compounds as morphological structure reactive templates reacted with the dissolved metal ions in the solution to form the bimetallic oxides. Their advantages are to extend the nanostructure of bimetallic oxides [35–41]. Herein, we briefly introduced the classic template methods and metal ion-reactive template methods used in the synthesis of nanostructured bimetallic oxides as anodes of LIBs in this chapter (**Figure 1**).

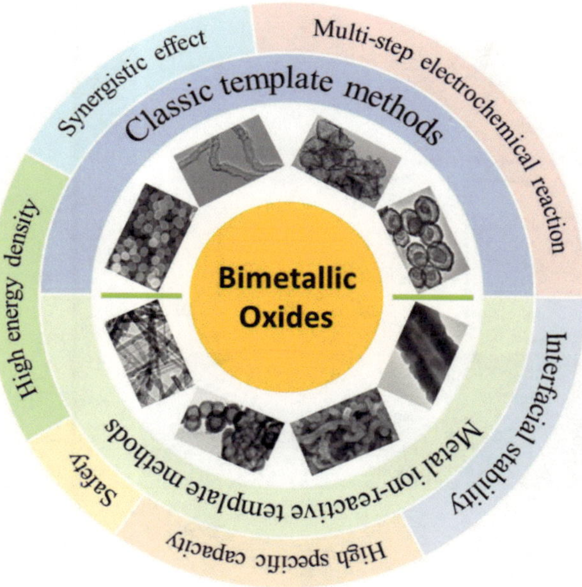

Figure 1.
Schematic diagram of the nanostructured bimetallic oxides synthesized by various template methods and their advantages used in LIBs.

2. Classic template methods

2.1 Amphiphilic polymer templates

Amphiphilic polymers as templates are used to absorb metal ions to control the morphological structures. In most instances, the hydrophilic functional group can absorb the metal ions for reaction and the hydrophobic functional group self-assembled as core to promote the formation of nanostructured bimetallic oxides [22–24]. In a report of Prof. Chen's group [22], shell-hybridized worm-like micelles (HPWs) were synthesized by using DNA/PEG_{113}-b-$P4VP_{74}$ core-cross-linked polymeric worm-like micelles (PWMs) as one-dimensional (1D) template, as shown in schematic illustration (**Figure 2A**). In the liquid phase reaction process, the outer-shell polyethylene glycol (PEG) chains coupled with Zn^{2+}/Co^{2+}, and the c-P4VP core with the positive charges was repulsive to Zn^{2+}/Co^{2+} to form a core-shell structure. After annealed at high temperature, the feeding mole ratio of Co^{2+}/Zn^{2+} is 2:1 (HPWs-2) were converted into hybrid carbon/$ZnCo_2O_4$ nanotubes (HCNTs-2). From transmission electron microscopy (TEM) image (**Figure 2B**), the product had a very good morphological structure stability by a soft template method. N_2 adsorption and desorption analysis showed that HCNTs-2 networks had a high Brunauer-Emmett-Teller (BET) specific surface area of 825 $m^2\ g^{-1}$ (**Figure 2C**) with a predominant peak at 33 nm and a satellite peak below 5 nm calculated by Barrett-Joyner-Halenda (BJH) model (**Figure 2D**). Energy-dispersive X-ray spectroscopy (EDS) elemental mapping analyses showed that the material had four evenly distributed elements of Zn, Co, O, and C (**Figure 2E**). After 600 cycles, the material could still release high capacities of 906 mAh g^{-1} and 494 mAh g^{-1} at the constant current densities of 1 A g^{-1} and 5 A g^{-1}, respectively. Such superior electrochemical performance was ascribed to an ingenious

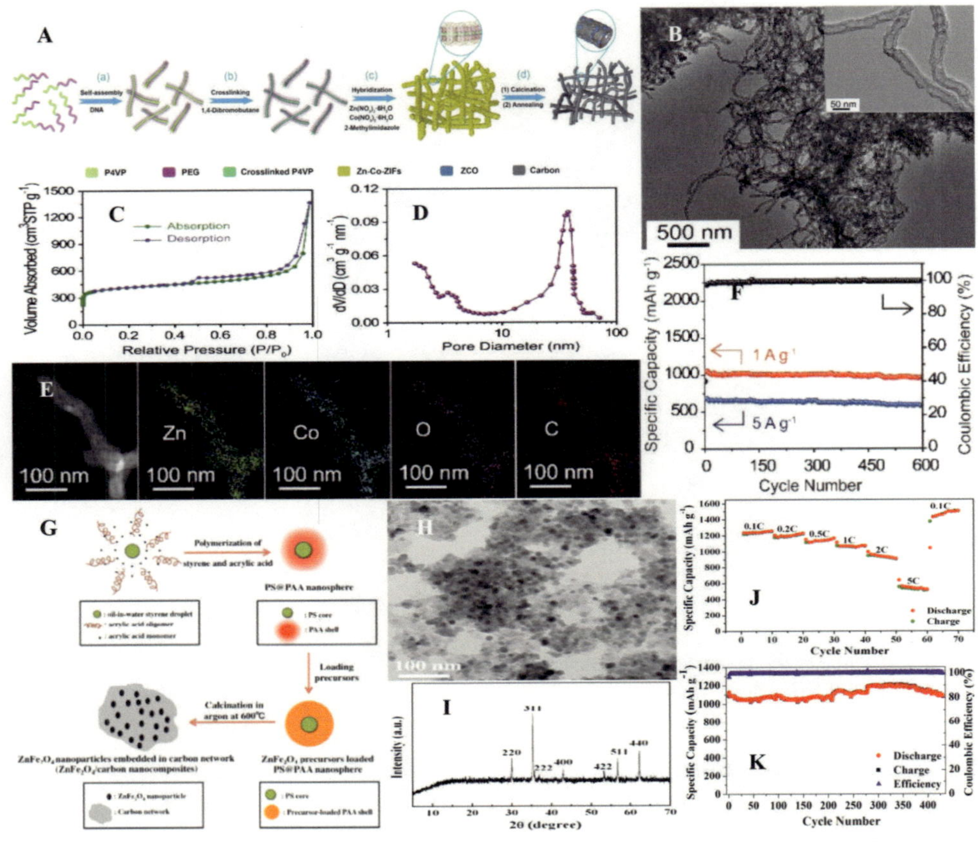

Figure 2.
PWs template: (A) Schematic illustration of the fabrication procedure of hybrid carbon/$ZnCo_2O_4$ nanotubes. (B) TEM image, (C) N_2 adsorption-desorption curves, (D) BJH peaks fitted, (E) EDS mapping analyses and (F) cycle performance of HCNTs derived from HPWs. (Ref. *[22]*, Copyright 2019 RSC) *PS@PAA nanospheres template: (G) Schematic diagram of the synthesis processes from monomers to $ZnFe_2O_4$/carbon nanocomposites. (H) TEM image, (I) XRD pattern, (J) rate performance and (K) cycle performance of the $ZnFe_2O_4$(79.3%)/carbon nanocomposites.* (Ref. *[23]*, Copyright 2016 ACS).

interaction carbon connections network structure with large amounts of opened micropores (**Figure 2F**).

In Prof. Lin's group [23], PS@PAA core-shell nanospheres as templates were synthesized by an emulsion polymerization reaction. **Figure 2G** shows a schematic illustration of the obtained $ZnFe_2O_4$/carbon nanocomposites. When pH value was controlled to 2, the carboxyl groups of the PAA shell coupled together with the metal moieties of precursors to form $ZnFe_2O_4$ precursor-loaded PS@PAA nanospheres with blurred edges by the strong coordination bonding. After thermal annealing under argon atmospheres, the precursor-loaded PS@PAA nanospheres were transferred into $ZnFe_2O_4$/carbon nanocomposites (**Figure 2H**), but the collapse, shrinkage, and carbonization of polymers resulted in an obvious change in the morphological structure. X-ray diffraction (XRD) pattern (**Figure 2I**) showed that the 2θ peaks at 30°, 35.5°, 38.0°, 43.5°, 53.2°, 56.7°, and 62.3° should be indexed to the planes of cubic spinel structure (*Fd-3 m*) $ZnFe_2O_4$ (Joint Committee on Powder Diffraction Standards (JCPDS) Card No. 022-1012). Owing to their unique structure, the attractive rate capacities of 1052 mAh g^{-1} (1 A g^{-1}), 926 mAh

g^{-1} (2 A g^{-1}), 521 mAh g^{-1} (5 A g^{-1}) and a stable cycle capacity of 1078 mAh g^{-1} (0.2 A g^{-1}) at 430 cycles can be observed in **Figure 2J** and **K**.

Zhang et al. [24] used polyvinyl pyrrolidone (PVP, Mw ~58,000) dissolved in ethylene glycol (EG), which might be formed as micro-emulsion and act as the soft template to absorb metal ions due to the strong coordination ability. After refluxing for 90 min at 170°C, ZnMn-glycolate hollow spheres were formed. The post-annealing treatment provided a driving force caused by a large contraction force (F_c) and an opposite adhesion force (F_a) to divide a single shell into two shells, thereby generating $ZnMn_2O_4$ with a ball-in-ball hollow structure. When used as anodes in LIBs, the delivered capacities were 490 mAh g^{-1} at the 50th cycle and 750 mAh g^{-1} at the 120th cycle under test conditions of a constant current density of 400 mA g^{-1} with a potential range from 0.01 V to 3.0 V, respectively. It is a classic phenomenon observed in Li-storage reaction process of transition metal oxides (TMOs), which should be attributed to possible electrode activation processes. When the current density reached 1200 mA g^{-1}, the discharge specific capacity was still kept at 396 mAh g^{-1}. The unique ball-in-ball hollow architecture could improve the structural integrity to withstand the mechanical strain caused by Li ions insertion/extraction, thereby improving the cycle stability. On the other hand, uniform and small primary nanoparticles could shorten the Li ions diffusion pathway, and the empty space could serve as an electrolyte reservoir to enhance the rate capability.

2.2 Polystyrenes templates

Polystyrenes (PS) spheres as templates are used to synthesize bimetallic oxides by different ways. Liu et al. [25] used sulfonated polystyrenes as templates to anchor Co-Fe precursor by a simple co-precipitation reaction, as shown in **Figure 3A**. By high-temperature calcination, the template was removed to form bowl-like $CoFe_2O_4$ hollow structure (**Figure 3B** and **C**). All the peaks can be successfully indexed to the crystal planes of $CoFe_2O_4$ crystalline structure (JCPDS Card No. 22-1086, Fd3m) (**Figure 3D**). When used in LIBs, the electrode still had a discharge capacity of about 1111 mAh g^{-1} after 200 cycles at 0.4 A g^{-1} (**Figure 3E**). Even at 1.2 A g^{-1}, the material delivered a reversible capacity of about 433 mAh g^{-1}. Most of the contributions came from the bowl-like structure that offered enough space to resist mechanical deformation caused by Li ions insertion and facilitated electrolyte infiltration for Li ions migration.

Prof. Kim et al. [26] designed colloidal PS templates by evaporating monodispersed PS spheres, which as rigid framework structures were soaked in a Zn/Fe precursor solution to fill the interstitial voids. A schematic diagram of the fabrication process of three-dimensionally ordered macroporous (3DOM) structure $ZnFe_2O_4$ (ZFO) crystals from (a) to (d) is illustrated in **Figure 3F**. Through air high-temperature sintering, the PS templates were removed to leave spinel $ZnFe_2O_4$ crystals with a 3DOM structure (**Figure 3G**). When used as anodes for LIBs, the ZFO@160 electrode delivered a high capacity of 1143.7 mAh g^{-1} over 100 cycles at 0.1 A g^{-1} (**Figure 3H**). Even at high current densities of 1.6 A g^{-1} and 3.2 A g^{-1}, the discharge capacities were still kept at 405.8 mAh g^{-1} and 286.4 mAh g^{-1} (**Figure 3I**), respectively, which were far higher than those of ZFO@55 and ZFO NPs. The stable 3DOM structure should facilitate electrolyte infiltration and buffering mechanical deformation.

Dr. Koo et al. [27] used PS spheres with the surface negative charge as templates to induce Zn^{2+}/Co^{2+} metal ions deposition on the templates' surface due to the electrostatic interaction. Through air calcination, the core PS spheres were removed to form

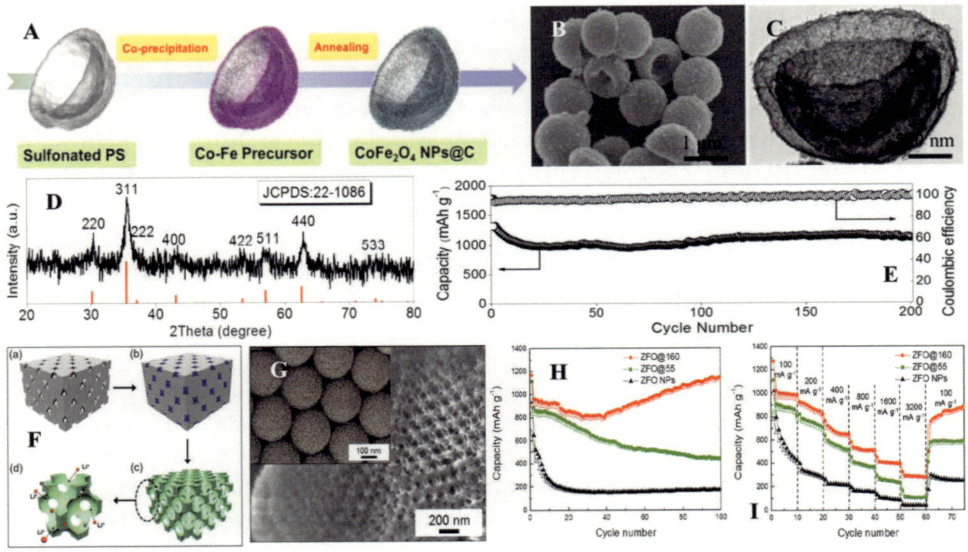

Figure 3.
Bool-like PS spheres template: (A) Schematic illustration of the synthetic procedure. (B) SEM image, (C) TEM image, (D) XRD pattern (E) Cycle performance (400 mA g^{-1}) of bowl-like $CoFe_2O_4$ NPs@C composites. (Ref. *[25]*, Copyright 2016 RSC) *Monodispersed PS spheres template: (F) Illustration of the overall fabrication process: (a) PS, (b) PS impregnated with zinc and iron, (c) 3DOM ZFO. (d) Schematic diagram of discharging process in 3DOM ZFO. (G) SEM image of 3DOM ZFO (the inset is colloidal PS templates). (H) Cycle performance and (I) Rate performance of ZFO@160, ZFO@55, and ZFO NPs.* (Ref. *[26]*, Copyright 2019 Elsevier).

porous $ZnCo_2O_4$ hollow spheres. As anodes in LIBs, their cycling capacity was obviously superior to that of $ZnCo_2O_4$ nanoparticles.

2.3 Bubble templates

The bubbles derived from the decomposing gases of reaction agents act as template to boost the formation of nanostructured bimetallic oxides. Dr. Ding et al. [28] reported a urea-assisted synthesis routine to synthesize Ni-Co oxide hollow microspheres as anodes in LIBs. The micro-bubbles (CO_2 and NH_3) formed by decomposition of urea molecule could serve as a soft template for deposition of Ni^{2+} and Co^{2+}. Then, the precursor rapidly nucleated around the micro-bubbles to boost a wrap of multi-layer of cations for the formation of hollow structure. Note that the gas flow led to the growth of needle-like structure. With the reaction time, the hollow spheres with well-ordered nanoneedles were synthesized under the dual action of morphological tailing and Ostwald ripening process. As anodes of LIBs, the material had superior capacities of 1080, 800, 580, and 220 mAh g^{-1} at various current densities of 100, 300, 500, and 4000 mA g^{-1}. The excellent electrochemical performance should be mainly derived from the unique and complex hollow porous structure with nanoneedles.

2.4 Inorganic templates

In general, many inorganic templates such as Cu_2O, carbon spheres and SiO_2 spheres are considered as hard templates that are widely used in the fabrication process of bimetallic oxides [29–33]. Guo et al. [29] reported that octahedral Cu_2O (**Figure 4A**) as a sacrificial template was used to construct hollow octahedral $ZnCo_2O_4$ nanocages assembled by ultrathin two-dimensional (2D) nanosheets (**Figure 4B** and **C**). High-resolution (HR)-TEM showed that the 0.284 nm and 0.243 nm lattice spacing were coincided with (220) and (311) planes of spinel structure $ZnCo_2O_4$, and the elemental mapping images demonstrated that the elements of O, Co, and Zn were evenly distributed in the hollow hierarchical products (**Figure 4D**). As shown in **Figure 4E**, the octahedral Cu_2O templates were etched by the reagent thiosulfate ($S_2O_3^{2-}$), while the OH^- ions generated in the etching reaction and hydrolysis of excess $S_2O_3^{2-}$ could drive the formation of $(ZnCo_x)O(OH)$ hollow nanocages. When etching time was 20 min, the obtained material had a uniform nanocage structure with a high BET specific area of 179.25 $m^2\ g^{-1}$ and average pore diameters of about 3.838 nm (**Figure 4F**). At 200 mA g^{-1}, the material can deliver a high reversible Li-storage capacity of 1025 mAh g^{-1} after 200 cycles, which had a more

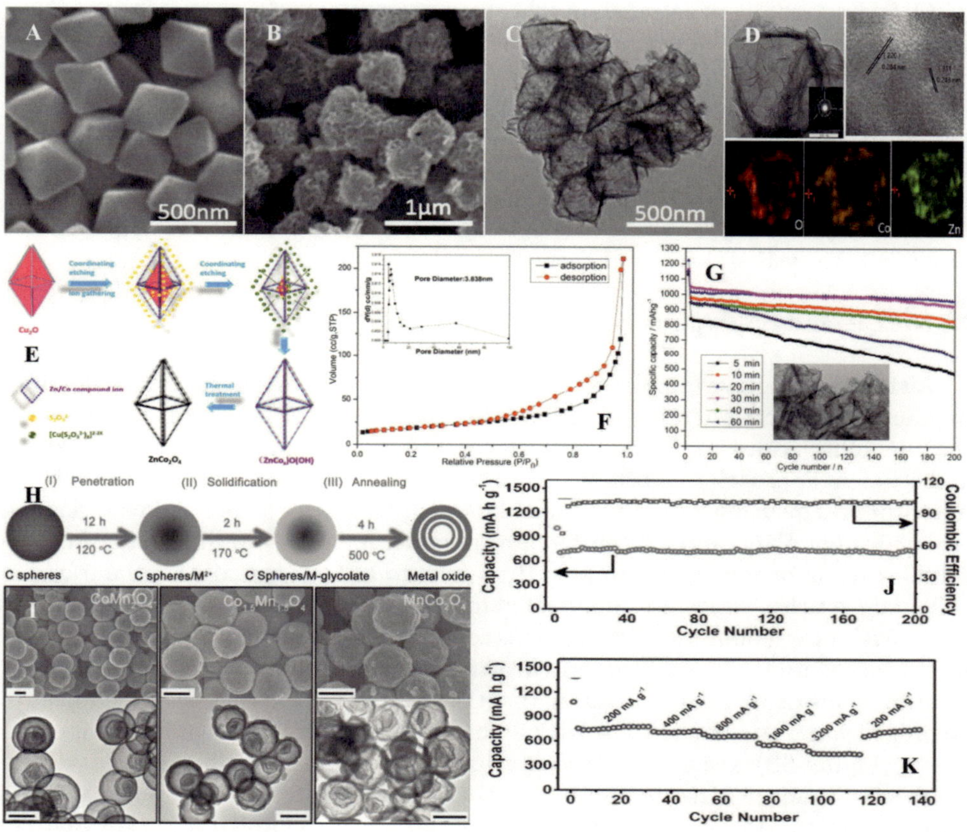

Figure 4.
Cu_2O template: SEM image (A) of Cu_2O octahedron. (B) SEM image of $(ZnCo_x)O(OH)$ precursors, (C) TEM image, (D) EDX mapping images, (E) Representative synthesis routine illustration, (F) Nitrogen adsorption/desorption isotherm and BJH pore size distribution plot (inset) of hollow octahedra $ZnCo_2O_4$ nanocages. (G) cycling performance at 200 mA g^{-1} of the different $ZnCo_2O_4$ materials obtained by varying with the coordinating etching reaction from 5 min to 60 min. (Ref. *[29]*, Copyright 2017 RSC) *Carbon spheres template: (H) Schematic illustration of the formation process of complex metal oxide multi-shelled hollow spheres. (I) FESEM and TEM images of $CoMn_2O_4$, $Co_{1.5}Mn_{1.5}O_4$ and $MnCo_2O_4$* (Scale bars: 500 nm). *(J) cycling performance and corresponding Coulombic efficiency at 200 mA g^{-1}. (K) rate performance at various current densities from 200 mA g^{-1} to 3200 mA g^{-1}.* (Ref. *[31]*, Copyright 2014 Wiley).

stable storage than those of other materials when etching in other different times (**Figure 4G**). By observation of TEM in the inset of **Figure 4G**, the material had still a well-kept structure without breakage after 200 cycles at 500 mA g^{-1}. These results demonstrated that the hollow structure shortened Li-ion diffusion path to improve the rate performance and offered a void space to alleviate the mechanical deformation for the cycling stability.

Hou et al. [30] also used octahedral Cu_2O as hard template and hydroxypropyl cellulose (HPC) as a film-forming agent to synthesize the $NiCo_2O_4$@CuO. During this process, the ions of Ni^{2+}, Co^{2+}, and OH- were aggregated on the surface of CuO octahedra with assistance of HPC, which gradually developed and then grew to form the Ni-Co-precursor subunits. After calcination, the inner CuO templates were removed to obtain hollow $NiCo_2O_4$ cages by immersion of a diluted ammonia solution. At a current density of 200 mA g^{-1}, the specific capacity was up to 793 mAh g^{-1} over 100 cycles. Even at 1400 mA g^{-1}, the average discharge capacities were still arrived at a high level of 377 mAh g^{-1}.

Metal ions readily penetrated into the interior porous structure of carbon spheres to form multi-layer bimetallic oxides with hollow-structure spheres, such as $CoMn_2O_4$, $Co_{1.5}Mn_{1.5}O_4$, $MnCo_2O_4$, $ZnMn_2O_4$, $NiCo_2O_4$, and $ZnCo_2O_4$ multi-shelled hollow spheres [31], and $ZnCo_2O_4$ and $ZnMn_2O_4$ yolk-shell spheres [32, 33]. In the report [31], a general "penetration-solidification-annealing" strategy was proposed to explain the phenomenon (**Figure 4G-I**). When tested as anodes in LIBs, carbon-coated $CoMn_2O_4$ triple-shelled hollow spheres had a high specific capacity of around 708.9 mAh g^{-1} during 200 cycles at 200 mA g^{-1}. The multi-shell hollow structure promoted the cycling stability of these materials.

2.5 N-CCPs and amphiphilic polymers mixed templates

Although the above templates have widely boosted the nanostructure formation in practical application, it is still difficult to control the morphological structure of the bimetallic oxides under conventional conditions. Herein, we designed a novel and efficient approach to synthesize porous and uniform Zn-Mn carbonates sub-microspheres by using a combined template with nanocarbon colloidal particles (inorganic template) and PVP surfactant (polymer template) [34]. Herein, the soluble carbon nanoparticles ranging from 1 nm to 100 nm are called as nano-carbon colloidal particles, which are abbreviated to N-CCPs. N-CCPs were obtained by electrolysis of ethanol in a LiOH aqueous solution [42]. A significant Tyndall effect was observed under a laser beam in a brownish red transparent N-CCPs aqueous solution (**Figure 5A**), demonstrating that the N-CCPs were evenly dispersed in deionized water as particles. The N-CCPs had a main diameter range from 6 nm to 20 nm (**Figure 5B**). Using Na_2CO_3 as precipitant, Zn-Mn carbonates spheres were synthesized under a synergistic effect of N-CCPs and PVPs surfactant in an aqueous solution, which had a uniform diameter of about 700 nm (**Figure 5C**). EDS analyses data showed that the material contained the elements of Zn, Mn, and O, and the Zn/Mn mole ratio was close to 1:2 (**Figure 5D**). After calcination, the obtained sample was still kept at a spherical structure (**Figure 5E**). TEM image (the inset of **Figure 5E**) showed that the sintered $ZnMn_2O_4$ spheres were uniform and hollow structure. For comparison, without the assistance of N-CCPs, the obtained Zn-Mn carbonates had a width diameter distribution from 0.75 μm to 3.2 μm (**Figure 5F**). TEM image (the inset of **Figure 5F**) showed that the sintered $ZnMn_2O_4$ spheres were solid and nonuniform structure. Therefore, a possible synthesis process of

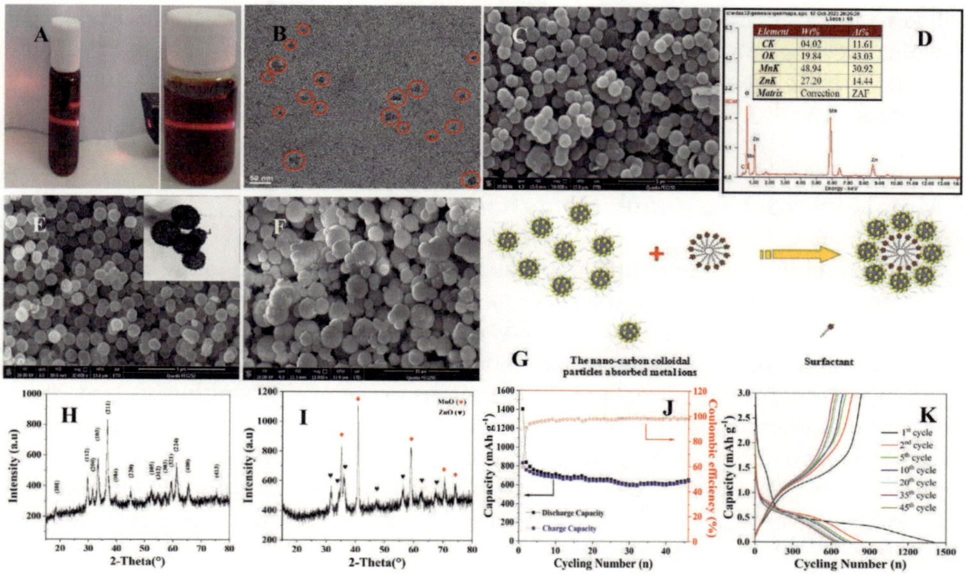

Figure 5.
*The combined template of N-CCPs and PVP: (A) Digital image of N-CCPs aqueous solution prepared by electrolysis; (B) TEM images of N-CCPs; (C) SEM image and (D) EDS analysis of Zn-Mn carbonates spheres prepared with N-CCPs; (E) SEM image of Zn-Mn bimetallic oxides prepared after calcination (Inset is TEM image of the sintered $ZnMn_2O_4$ hollow spheres); (F) SEM image of non-uniform Zn-Mn carbonates spheres prepared without N-CCPs (Inset is TEM image of the sintered $ZnMn_2O_4$ solid spheres); (G) The schematic diagram of possible synthesis mechanism of uniform spherical Zn-Mn carbonates by an assembly reaction. XRD patterns of Zn-Mn bimetal oxides sub-microspheres prepared by N-CCPs in (H) air and (I) argon calcination, respectively. (J) The electrochemical cycling performance and (K) cycling curves of Zn-Mn bimetallic oxides@N-CCPs sub-microspheres in argon calcination. (*Ref. [34] Invention Patent of P. R. China, ZL. 202210389708.5 [P]*).*

uniform Zn-Mn compounds was described here (**Figure 5G**). N-CCPs containing a great amount of oxygen-rich functional groups had an excellent chemical activity, which can absorb metal ions to form a stable community in an aqueous solution due to strong electrostatic interactions, as the previous reports [43, 44]. Subsequently, using N-CCPs as ion carriers (hard template) of precursors to prompt self-assembly reaction, the surface functional groups of N-CCPs and metal ions would easily couple with amphiphilic high polymer (PVP, K30, soft template) due to the coordination bond and other covalent bonds in this reaction system. In this case, the surface of N-CCPs with abundant oxygen-containing functional groups served as nucleation sites for the stable and rapid growth of Zn-Mn carbonates nuclei to form a uniform spherical structure [43, 44]. Therefore, it is realized that it is a new type of nano-composite assembly mode, which breaks through the size and structure limits of the traditional hard template, and makes full use of the dynamic balance cavity provided by the soft template. Therefore, the feasibility and stability of the reaction were improved by this way. When annealed at air atmospheres, the obtained sample possessed a very good crystallinity of spinel $ZnMn_2O_4$ crystals (**Figure 5H**). However, when annealed at argon atmospheres, the obtained sample had cubic phase MnO crystals and hexagonal ZnO crystals (**Figure 5I**). The Zn-Mn bimetallic oxides@N-CCPs were fabricated as the disc electrode with 1.5 cm in diameter, which was used to measure the electrochemical performance. When the current density was 100 mA g^{-1}, the first discharge and charge capacities were 1406.4 mAh g^{-1} and 831.92 mAh g^{-1}, respectively (**Figure 5J**). After 45 cycles, the delivered capacity was still kept at 650.94 mAh g^{-1} with a high Coulombic efficiency of 97.9%. The electrochemical cycling curves showed that the first discharge potential platform was about 0.5 V, and the following discharge potential platforms were stabilized at about 0.7 V (**Figure 5K**).

3. Metal ion-reactive template methods

The metal ion-reactive template methods have been extended to the synthesis of various bimetallic oxides in the past two decades [35–41]. Normally, the readily synthesized metal compounds were used as morphological structure reactive templates and metal ion sources to prepare ternary metal oxides. In liquid phase chemical reaction system, the functional groups or cations in the readily synthesized metal compounds were replaced by the exchanged metal ions. Using solid ZnO nanorod arrays as a reactive template, $ZnMn_2O_4$ mesoscale tubular arrays were successfully synthesized via the cation exchange method in a manganese acetate aqueous solution. XRD analyses were used to observe the time-dependent evolution process on crystal structures [35]. As the reaction time was increased to 48 h at room temperature, the high purity $ZnMn_2O_4$ phase was obtained. As anodes for LIBs, $ZnMn_2O_4$ tubular arrays delivered higher average capacities of 644.6, 578.8, 515.9, 437.5, 363.6, and 243.5 mAh g^{-1} at the current densities of 100, 200, 400, 800, 1600, and 3200 mA g^{-1}, which were obviously superior to those of $ZnMn_2O_4$ nanoparticles.

Ternary $Zn_2Ti_3O_8$ and $ZnMn_2O_4$ crystals were successfully synthesized in an alkaline Zn^{2+} ions aqueous solution by a simple ion exchange reaction [36–38]. In an alkaline environment, the dissociation zinc ions in solution were easily diffused into the solid lattice of the metal compounds as template. For example, in an ammonia solution at 60°C, Zn^{2+} ions as exchanged ions were used to replace Na^+ and H^+ cations of $Na_xH_{2-x}Ti_3O_7$ nanotubes as structure templates for the synthesis of $Zn_2Ti_3O_8$ nanowires [36]. Ternary $Zn_2Ti_3O_8$ crystal was composed of octahedral [TiO_6] and tetrahedral [ZnO_4], which could provide a diffusion space for Li ions insertion/extraction in LIBs. Zhang et al. [37, 38] reported that $NaBH_4$ dissolved into a NaOH aqueous solution as a reductant reacted with MnO_2 for the formation of Mn_3O_4. Subsequently, Zn^{2+} ions can gradually diffuse into the solid lattice of Mn_3O_4. In this process, the manganese vacancies were occupied by Zn^{2+} ions, or manganese ions partially substituted by Zn^{2+} ions in the tetrahedral or octahedral interspaces. Under constant stirring, the Zn-Mn-O solid solution recrystallized and gradually converted into the final $ZnMn_2O_4$ phase at room temperature. By XRD and EDS mapping analyses, the 2θ diffraction peaks were completely indexed to the planes of spinel structure $ZnMn_2O_4$, [37, 38] and zinc was even distributed in $ZnMn_2O_4$ nanotubes as well as manganese [38]. A possible reduction/ion exchange/recrystallization mechanism was proposed to explain the reaction process.

In recent years, we provided a simple alcohol solvothermal zinc ion-reactive template method to synthesize $Zn_2Ti_3O_8$ nanocrystals [39–41]. Using alcohol as the solvent, Zn^{2+} ions were adsorbed and then reacted with amorphous titanium compounds to obtain fluffy-shaped $Zn_2Ti_3O_8$ crystals by stirring at 60°C and then solvothermal reaction. There are some advantages by using ethanol as solvent instead of water. First, the surface tension of ethanol (22.3 mN, 20°C) is significantly lower than that of water (72.8 mN, 20°C). The dissolved metal ions in ethanol are conductive to diffusion into the inner metal compounds as the morphological templates, forming an effective penetration effect. Second, many metal compounds precursors contain abundant organic functional groups that may inhibit partial water into the inner metal compounds. Ethanol has the characteristics of amphoteric substances, which are both hydrophilic (hydroxyl groups) and hydrophobic (carbon chains). It is more conducive to improving metal ions to break through the bondage of surface functional groups of reactive template. Third, the hydrated zinc ions ($[Zn(H_2O)_6]^{2+}$) in aqueous solution have a bigger size, resulting in a poor diffusion ability [45]. In the ethanol solvent, Zn^{2+} has little tendency to form the hydrated metal ions, and thus it has a higher ability to diffuse into the inner metal compounds. Therefore, it is easy to synthesize bimetallic oxides with good crystallinity by a simple metal ion exchange reaction when using ethanol as solvent.

A simple schematic diagram of the possible synthesis process of $Zn_2Ti_3O_8$ crystals is depicted in **Figure 6A**. Based on this strategy, $Zn_2Ti_3O_8$ hollow spheres with different shell thicknesses can be obtained by controlling the shell thickness of the SiO_2@ TiO_2 core-shell spheres [39]. The moisture derived from OH- or glycolate reacting with CH_3COO- and carried between TiO_2 molecules could contribute to forming an acidic environment that could drive a very complete Zn-Ti-O dissolution-regeneration reaction. Nanostructured $Zn_2Ti_3O_8$ crystals were synthesized using titanium compounds as reactive templates and zinc ions as exchanged ions. Notably, too much moisture would result in the structural changes in the final products, [40] and a lack of moisture would result in an incomplete zinc ion exchange reaction. Therefore, moisture played a very important role in controlling the zinc ion exchange reaction process [40]. By the observation of TEM images (**Figure 6C** and **D**), $Zn_2Ti_3O_8$ solid spheres had a uniform flocculent inner structure, which was obviously different from that of TiO_2 solid spheres. Three elements of Zn, Ti, and O were uniformly distributed in (**Figure 6E**) in the material by EDS mapping images. XRD patterns (**Figure 6F**) also demonstrated that the obtained samples had a very obvious crystal

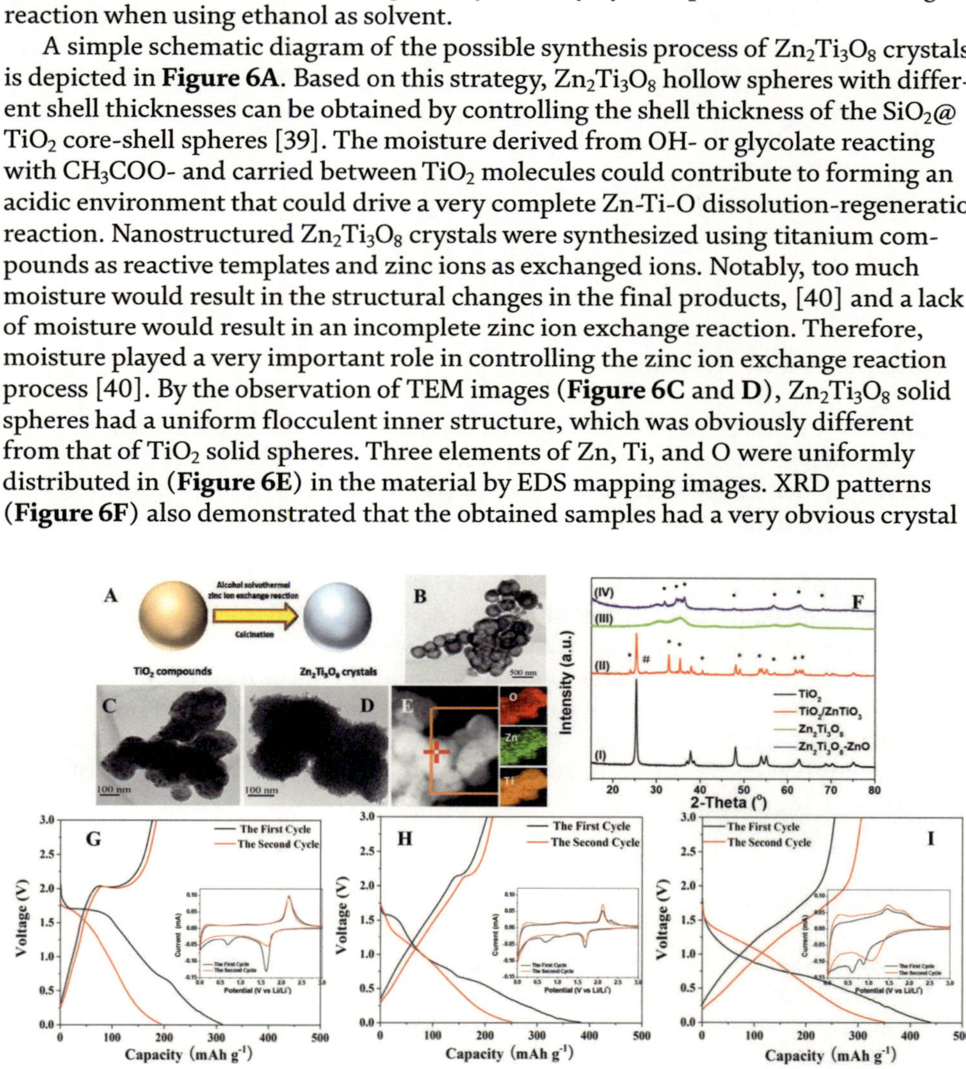

Figure 6.
Titanium compounds templates: (A) Schematic diagram of the possible synthesis process of $Zn_2Ti_3O_8$ crystals during the alcohol solvothermal treatment. (B) TEM images of $Zn_2Ti_3O_8$ hollow spheres. TEM images of (C) TiO_2 and (D) $Zn_2Ti_3O_8$ nanocrystals. (E) EDS mappings of $Zn_2Ti_3O_8$ nanocrystals. (F) XRD patterns of (I) TiO_2, (II) $TiO_2/ZnTiO_3$ and (III) $Zn_2Ti_3O_8$ solid spheres and (IV) $Zn_2Ti_3O_8$-ZnO composites. The voltage profiles (current density: 300 mA g^{-1}) and CV curves (insets, scan rate: 0.1 mV s^{-1}) of (G) TiO_2, (H) $TiO_2/ZnTiO_3$, and (I) $Zn_2Ti_3O_8$ solid spheres in the potential range from 0.01 to 3.0 V. (Ref. *[39–41]*, Copyright 2016 & 2019 Elsevier & 2023 Springer).

structure change by varying with the Zn/Ti atomic ratios. In the absence of zinc ions, the obtained sample had an anatase phase crystal structure after calcination at 500°C. At a 1:3 Zn/Ti atomic ratio, the sample contained anatase TiO_2, ecandrewsite $ZnTiO_3$, and a small amount of rutile TiO_2. At a 2:3 Zn/Ti mole ratio, the high-quality cubic phase $Zn_2Ti_3O_8$ crystals were obtained. With Zn/Ti mole ratio further increased to 3:3, there were cubic phase $Zn_2Ti_3O_8$ solid spheres and hexagonal ZnO nanoparticles in the product [40, 41].

By controlling the Zn/Ti mole ratios to 0:3, 1:3, and 2:3, the delivered capacities of the samples could achieve 311.5, 382.8, and 439 mAh g^{-1} at the first cycles, respectively (**Figure 6G-I**). By the given capacities and the mole mass of materials, the amounts of Li ions insertion can be precisely calculated using formula (4). Compared to TiO_2 solid spheres, the decreased discharge voltage platform of $Zn_2Ti_3O_8$ solid spheres would lead to an improved energy density when applied in practical LIBs. The phenomena were also observed by Cyclic voltammetries (CV) tests (the insets of **Figure 6G-I**). Therefore, by this synthetic method, the effect of exchanged metal ions on the electrochemical reaction was precisely explored.

$$C = F \times n \times 1000 \tag{4}$$

4. Conclusion and outlooks

Bimetallic oxides, as anode materials of LIBs, have attracted much attention because of their low cost, high specific capacity, environmental benignity, and safety. To achieve optimal Li-storage electrochemical performance, building a porous and uniform morphological structure is an effective means. In this chapter, we briefly introduced the classic template methods and metal ion-reactive template methods used in the synthesis of bimetallic oxides, and elaborated their electrochemical performance and operation mechanism. Even so, there are still some problems worthy of further studying the nano-synthesis reaction. Herein, we proposed a combined template (N-CCPs and surfactants) method to synthesize Zn-Mn bimetallic oxides, used alcohol instead of water as solvent to synthesize bimetallic oxides in metal ion-reactive template methods, and discussed their Li-storage electrochemical mechanism. It is expected that these technologies can be used in the synthesis of other materials for various fields.

Acknowledgements

This work was funded by the Natural Science Foundation of Jiangxi, China (20202BABL203003), the Science and Technology Project of Education Department of Jiangxi, China (GJJ201017), and the Key Science and Technology Project of Ji'an City (2020-2103), China. The author sincerely acknowledges the Elsevier, RSC, Wiley, and ACS publishers.

Author details

Wenming Liao
School of Chemistry and Chemical Engineering, Jinggangshan University, China

*Address all correspondence to: liaowenming21@163.com

References

[1] Mukherjee R, Thomas AV, Krishnamurthy A. Photothermally reduced graphene as high-power anodes for lithium ion batteries. ACS Nano. 2012;**6**:7867. DOI: 10.1021/nn303145j

[2] Song M, Park S, Alamgir FM, et al. Nanostructured electrodes for lithium-ion and lithium-air batteries: The latest developments, challenges, and perspectives. Materials Science &Engineering R. 2011;**72**:203. DOI: 10.1016/j.mser.2011.06.001

[3] Kim TH, Song WT, Son DY, et al. Lithium-ion batteries: Outlook on present, future, and hybridized technologies. Journal of Materials Chemistry A. 2019;**7**:2942. DOI: 10.1039/C8TA10513H

[4] Zheng Y, Yao YZ, Ou JH, et al. A review of composite solid-state electrolytes for lithium batteries: Fundamentals, key materials and advanced structures. Chemical Society Reviews. 2020;**49**:8790. DOI: 10.1039/D0CS00305K

[5] Qi W, Shapter JG, Wu Q, et al. Nanostructured anode materials for lithium-ion batteries: Principle, recent progress and future perspective. Journal of Materials Chemistry A. 2017;**5**:19521. DOI: 10.1039/C7TA05283A

[6] Cha H, Kim J, Lee Y, et al. Issues and challenges facing flexible lithium ion batteries for practical application. Small. 2017;**14**:1702989. DOI: 10.1002/smll.201702989

[7] Wu FX, Maier J, Yu Y. Guidelines and trends for next generation rechargeable lithium and lithium-ion batteries. Chemical Society Reviews. 2020;**49**:1569. DOI: 10.1039/C7CS00863E

[8] Fan ZJ, Yan J, Wei T, et al. Nanographene-constructed carbon nanofibers grown on graphene sheets by chemical vapor deposition: High-performance anode materials for lithium ion batteries. ACS Nano. 2011;**5**:2787. DOI: 10.1021/nn200195k

[9] Zhong SY, Liu HZ, Wei DH, et al. Long-aspect-ratio N-rich carbon nanotubes as anode material for sodium and lithium ion batteries. Chemical Engineering Journal. 2020;**395**:125054. DOI: 10.1016/j.cej.2020.125054

[10] Lee M, Reddi RKR, Choi JB, et al. In operando AFM characterization of mechanical property evolution of Si anode binders in liquid electrolyte. ACS Applied Energy Materials. 2020;**3**:1899. DOI: 10.1021/acsaem.9b02332

[11] Stephenson T, Li Z, Olsen B, et al. Lithium ion battery applications of molybdenum disulfide (MoS_2) nanocomposites. Energy & Environmental Science. 2014;**7**:209. DOI: 10.1039/C3EE42591F

[12] Han F, Zhang CZ, Yang JX, et al. Well-dispersed and porous FeP@C nanoplates with stable and ultrafast lithium storage performance through conversion reaction mechanism. Journal of Materials Chemistry A. 2016;**4**:12781. DOI: 10.1039/C6TA04521A

[13] Wang ZY, Wang ZC, Liu WT, et al. Amorphous $CoSnO_3$@C nanoboxes with superior lithium storage capability. Energy & Environmental Science. 2013;**6**:87. DOI: 10.1039/C2EE23330D

[14] Fan BB, Hu AP, Chen XH, et al. Hierarchical porous $ZnMn_2O_4$ microspheres as a high-performance anode for lithium ion batteries.

Electrochimica Acta. 2016;**213**:37. DOI: 10.1016/j.electacta.2016.07.030

[15] Chen JX, Liu W, Liu S, et al. Marine microalgaes-derived porous $ZnMn_2O_4$/C microspheres and performance evaluation as Li-ion battery anode by using different binders. Chemical Engineering Journal. 2017;**308**:1200. DOI: 10.1016/j.cej.2016.09.144

[16] Cao HL, Zhou XF, Deng W, et al. Layer structured graphene/porous $ZnCo_2O_4$ composite film for high performance flexible lithium-ion batteries. Chemical Engineering Journal. 2018;**343**:654. DOI: 10.1016/j.cej.2018.03.001

[17] Ma YT, Xie QS, Liu X, et al. Synthesis of amorphous $ZnSnO_3$ double-shell hollow microcubes as advanced anode materials for lithium ion batteries. Electrochimica Acta. 2015;**182**:327. DOI: 10.1016/j.electacta.2015.09.102

[18] Yao LM, Hou XH, Hu SJ, et al. An excellent performance anode of $ZnFe_2O_4$/fake graphite composite for lithium ion battery. Journal of Alloys and Compounds. 2014;**585**:398. DOI: 10.1016/j.jallcom.2013.09.066

[19] Liu YR, Bai J, Ma XJ, Li JF, et al. Formation of quasi-mesocrystals $ZnMn_2O_4$ twin microspheres via an oriented attachment for lithium-ion batteries. Journal of Materials Chemistry A. 2014;**2**:14236. DOI: 10.1039/C4TA02950J

[20] Zhu SQ, Chen QL, Yang C, et al. Biomorphic template-engaged strategy towards porous zinc manganate micro-belts as a competitive anode for rechargeable lithium-ion batteries. International Journal of Hydrogen Energy. 2017;**42**:14154. DOI: 10.1016/j.ijhydene.2017.04.197

[21] Zhao CH, Shen Y, Qiu SE, et al. Hierarchical porous $ZnMn_2O_4$ derived from cotton substance as high-performance lithium ion battery anode. Micro & Nano Letters. 2016;**11**:287. DOI: 10.1049/mnl.2016.0086

[22] Li HY, Lv LL, Wang WC, et al. A network of porous carbon/$ZnCo_2O_4$ nanotubes derived from the shell-hybridized worm-like micelles for lithium storage. Journal of Materials Chemistry A. 2019;**7**:22642. DOI: 10.1039/C9TA07869J

[23] Jiang BB, Han CP, Li B, et al. In-situ crafting of $ZnFe_2O_4$ nanoparticles impregnated within continuous carbon network as advanced anode materials. ACS Nano. 2016;**2**:2728. DOI: 10.1021/acsnano.5b07806

[24] Zhang GQ, Yu L, Wu HB, et al. Formation of $ZnMn_2O_4$ ball-in-ball hollow microspheres as a high-performance anode for lithium-ion batteries. Advanced Materials. 2012;**24**:4609. DOI: 10.1002/adma.201201779

[25] Liu JX, Liang J, Zhou H, et al. $CoFe_2O_4$ nanoparticles anchored on bowl-like carbon backbone for enhanced reversible lithium storage. RSC Advances. 2016;**6**:50153. DOI: 10.1039/C6RA09160A

[26] Kim JG, Noh Y, Kim YM, et al. Formation of ordered macroporous $ZnFe_2O_4$ anode materials for highly reversible lithium storage. Chemical Engineering Journal. 2019;**372**:363. DOI: 10.1016/j.cej.2019.04.151

[27] Koo WT, Jang HY, Kim C, et al. MOF derived $ZnCo_2O_4$ porous hollow spheres functionalized with Ag nanoparticles for a long-cycle and high-capacity lithium ion battery anode. Journal of Materials Chemistry A. 2017;**5**:22171. DOI: 10.1039/C7TA07573A

[28] Ding CH, Yan D, Zhao YJ, et al. A bubble-template approach for assembling Ni-Co oxide hollow microspheres with an enhanced electrochemical performance as an anode for lithium ion batteries. Physical Chemistry Chemical Physics. 2016;**18**:25879. DOI: 10.1039/C6CP04097G

[29] Liu BH, Liu H, Liang MF, et al. Controlled synthesis of hollow octahedral $ZnCo_2O_4$ nanocages assembled by ultrathin 2D nanosheets for enhanced lithium storage. Nanoscale. 2017;**9**:17174. DOI: 10.1039/C7NR06259A

[30] Hou XY, Xue S, Liu MT, et al. Hollow irregular octahedra-like $NiCo_2O_4$ cages composed of mesoporous nanosheets as a superior anode material for lithium-ion batteries. Chemical Engineering Journal. 2018;**350**:29. DOI: 10.1016/j.cej.2018.05.164

[31] Zhang GQ, Lou XW. General synthesis of multi-shelled mixed metal oxide hollow spheres with superior lithium storage properties. Angewandte Chemie, International Edition. 2014;**126**:9187. DOI: 10.1002/anie.201404604

[32] Deng JJ, Yu XL, Qin XY, et al. Carbon sphere-templated synthesis of porous yolk-shell $ZnCo_2O_4$ spheres for high performance lithium storage. Journal of Alloys and Compounds. 2019;**780**:65. DOI: 10.1016/j.jallcom.2018.11.331

[33] Zhang LX, Wang YL, Jiu HF, et al. Hollow core-shell $ZnMn_2O_4$ microspheres as a high-performance anode material for lithium-ion batteries. Ceramics International. 2015;**41**:9655. DOI: 10.1016/j.ceramint.2015.04.031

[34] Liao WM, Chen HH, Qiu TQ et al. A fabrication method of spherical Zn-Mn metal compound. Invention Patent of P. R. China, ZL. 202210389708.5. 2023-04-11

[35] Kim JG, Lee SH, Kim Y, et al. Fabrication of free-standing $ZnMn_2O_4$ mesoscale tubular arrays for lithium-ion anodes with highly reversible lithium storage properties. ACS Applied Materials & Interfaces. 2013;**5**:11321. DOI: 10.1021/am403546s

[36] Hong ZS, Wei MD, Deng QX, et al. A new anode material made of $Zn_2Ti_3O_8$ nanowires: Synthesis and electrochemical properties. Chemical Communications. 2010;**46**:740. DOI: 10.1039/B916681E

[37] Yuan CZ, Zhang LH, Hou LR, et al. Scalable room-temperature synthesis of mesoporous nanocrystalline $ZnMn_2O_4$ with enhanced lithium storage properties for lithium-ion batteries. Chemistry–A European Journal. 2015;**21**:1262. DOI: 10.1002/chem.201404624

[38] Zhang LH, Zhu SQ, Cao H, et al. Hierarchical porous $ZnMn_2O_4$ hollow nanotubes with enhanced lithium storage toward lithium-ion batteries. Chemistry–A European Journal. 2015;**21**:10771. DOI: 10.1002/chem.201501421

[39] Liao WM, Tian JH, Shan ZQ, et al. Facile synthesis of $Zn_2Ti_3O_8$ hollow spheres based on ion exchange as promising anodes for lithium ion batteries. Electrochimica Acta. 2016;**216**:94. DOI: 10.1016/j.electacta.2016.08.124

[40] Liao WM, Li WF, Tian JH, et al. Solvothermal ion exchange synthesis of ternary cubic phase $Zn_2Ti_3O_8$ solid spheres as superior anodes for lithium ion batteries. Electrochimica Acta. 2019;**302**:363. DOI: 10.1016/j.electacta.2019.02.014

[41] Liao WM, Chen HH, Zeng YY, Liu LM. Recent progress in the fabrication of nanostructured zinc-based ternary metal oxides for high-performance lithium-ion batteries. Journal of Applied Electrochemistry. 2023;**53**:1077. DOI: 10.1007/s10800-022-01832-5

[42] Jing MJ, Wang JF, Hou HS, et al. Carbon quantum dots coated Mn_3O_4 with enhanced performances for Lithium-ion batteries. Journal of Materials Chemistry A. 2015;**3**:16824. DOI: 10.1039/C5TA03610K

[43] Li YZ, Meng YS, Liu XL, et al. Double-protected zinc ferrite nanospheres as high rate and stable anode materials for lithium ion batteries. Journal of Power Sources. 2019;**442**:227256. DOI: 10.1016/j.jpowsour.2019.227256

[44] Chen RZ, Hu Y, Shen Z, et al. Controlled synthesis of carbon nanofibers anchored with $ZnxCo_{3-x}O_4$ nanocubes as binder-free anode materials for lithium-ion batteries. ACS Applied Materials & Interfaces. 2016;**8**:2591. DOI: 10.1021/acsami.5b10340

[45] Wang JY, Yang NL, Tang HJ, et al. Accurate control of multi-shelled Co_3O_4 hollow microspheres for high-performance anode materials in lithium ion batteries. Angewandte Chemie, International Edition. 2013;**52**(25):6417. DOI: 10.1002/anie.201301622